From Plant to Pigment
How to make your own vibrant inks, pastels, and paints

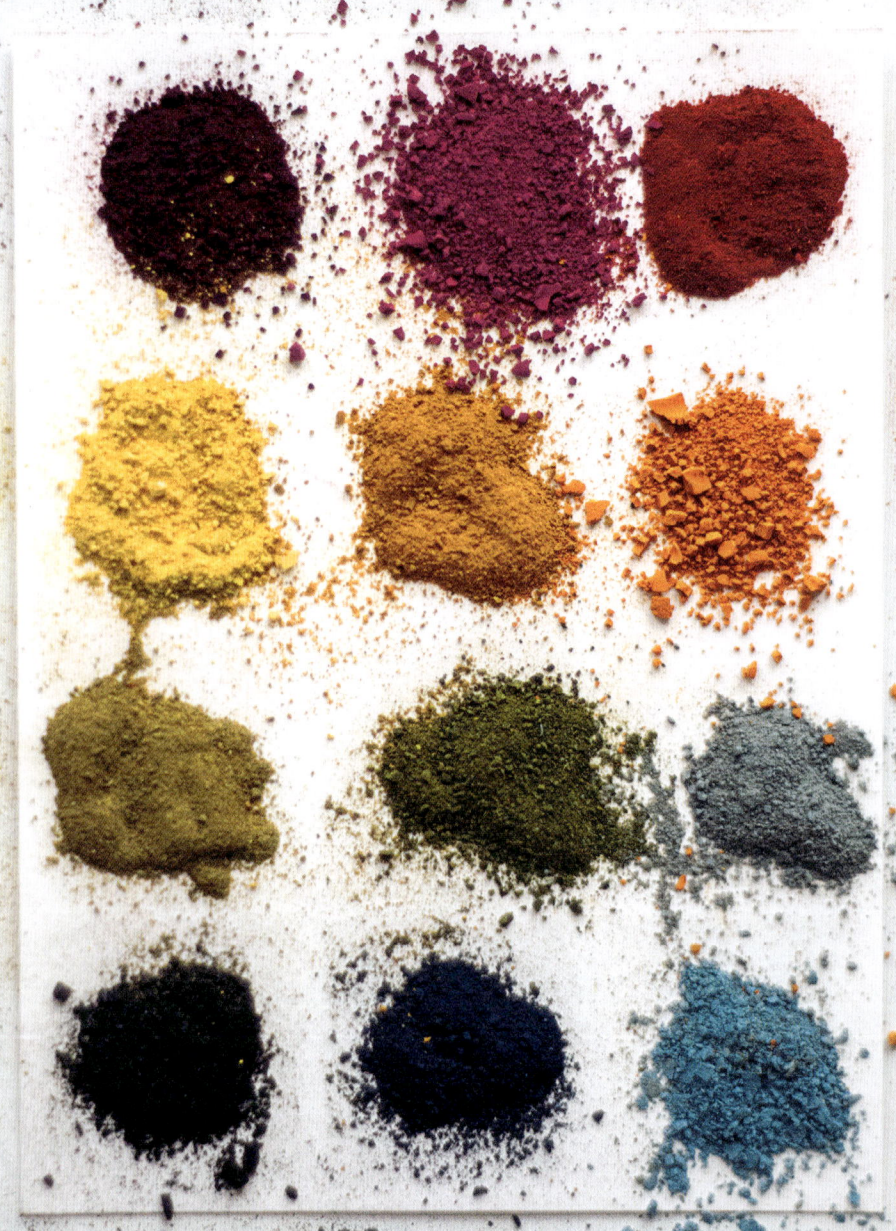

From Plant to Pigment
How to make your own vibrant inks, pastels, and paints

NATALIE STOPKA

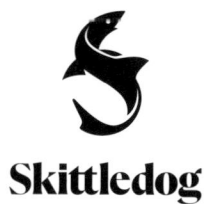

Skittledog

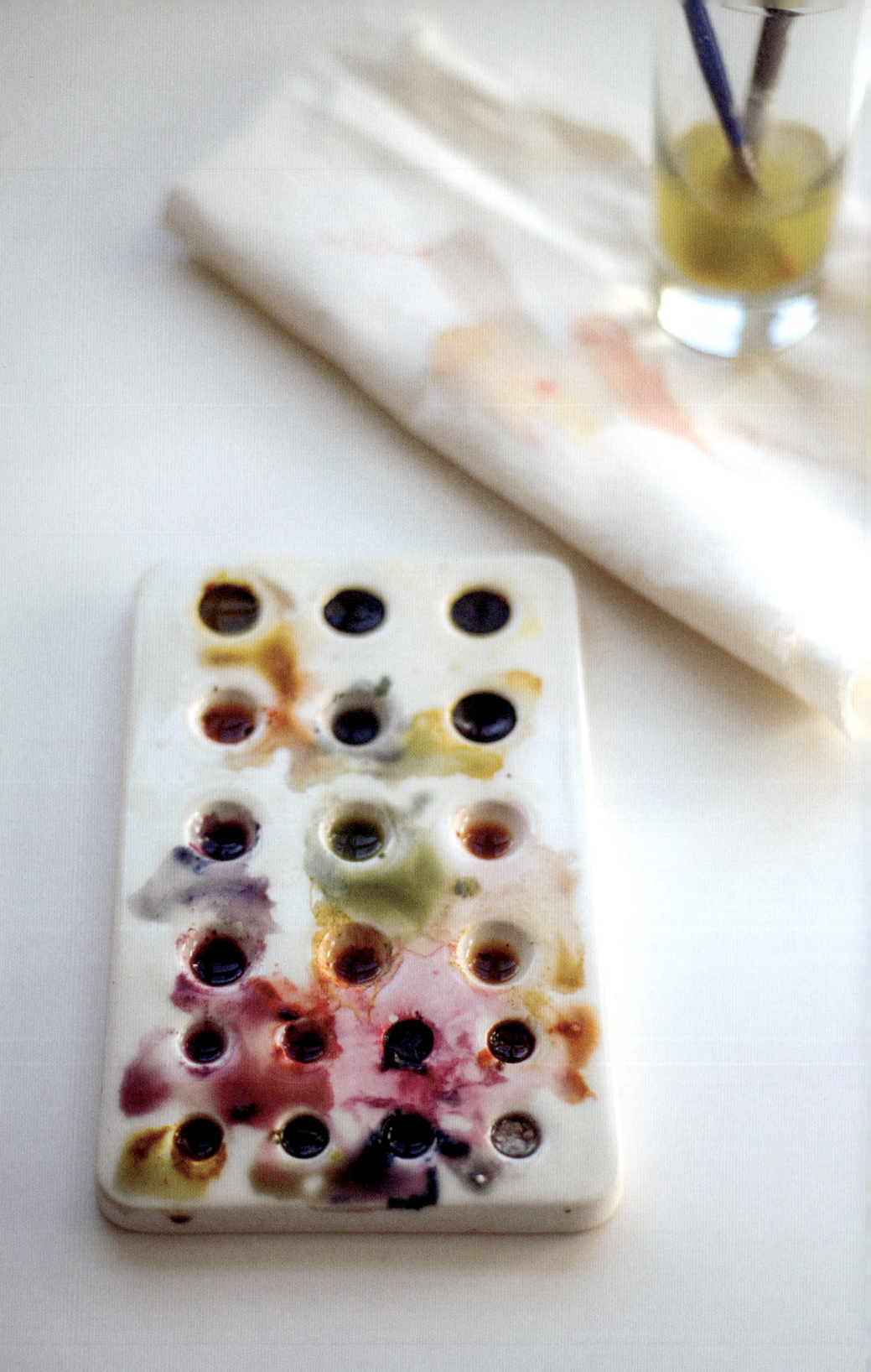

Contents

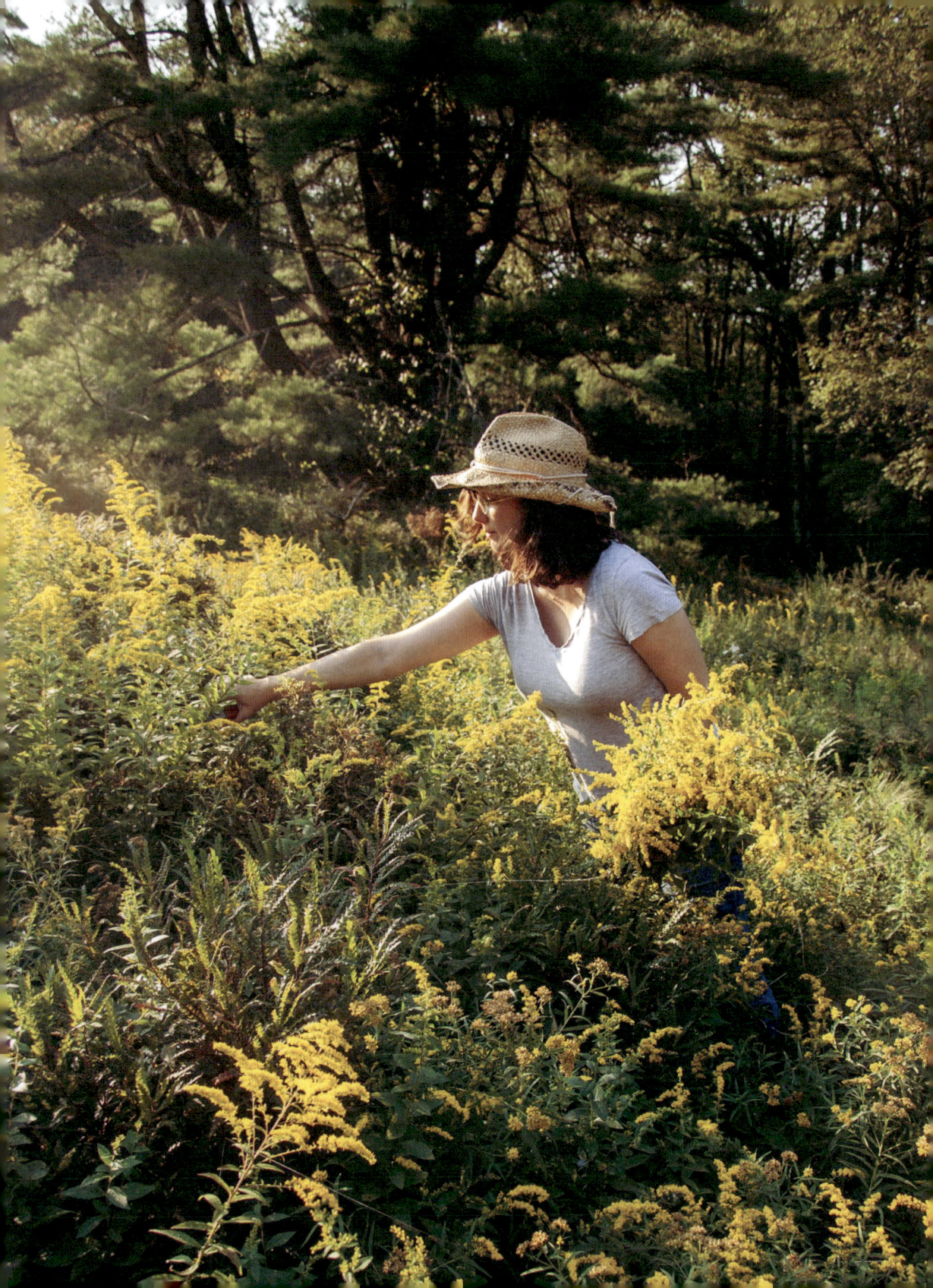

Introduction

The many plants that surround and sustain us are rich with color. With a little alchemical know-how, you can transform the brilliant hues of our botanical companions from liquid dye to granular pigment, supplying myriad artistic uses as paint, ink, and pastel.

The subtle and expressive colors of handmade botanical pigments not only visually evoke the natural world, but they also materially connect your artwork to nature. So, how and why do plants have the capacity to produce color? For one, flora and fauna, though intricately linked, subscribe to different modes of being. Rooted in place, plants—sprawling and climbing to harvest all the light they possibly can—fulfill their nutritional requirements from the sun above, the earth below, and the water flowing past. Their ability to do this makes them ingenious chemical engineers. Absorbed from the air and sucked from the soil are the elemental building blocks that a plant uses to cobble together a wide array of complex molecules.

To reproduce, some plants deploy colorful compounds in their flowers and fruit, visually signaling to birds and bees in a mutually beneficial cross-species collaboration. Other molecules serve as toxic defense mechanisms, protecting against predation by microbes, insects, and herbivores.

Whether created for plant communication, immune support, defense, or some still unknown purposes, we value the compounds that reflect to our eyes a beautiful spectrum of light. We perceive these biochromes as colors, and we utilize them as botanical dyes.

Over several thousand years, humans have developed cultural connections to botanical colors in a number of ways, including through ecological stewardship, plant husbandry, chemistry, and artistry. Through the ages, the skills of manipulating and applying botanical dyes have been refined, complexified, and specialized. In this book, I focus on a particular offshoot of this craft: the transmutation of ephemeral organic dyes into durable pigments for use by artists across various media.

Those who work with any sort of paint or pastel are likely familiar with a coterie of colors that are different from plant dyes. Artists' pigments are finely ground bits of colorful matter, prized for their steadfast, permanent hues. For example, natural pigments ground from minerals, rocks, and clay possess colors stabilized over eons. They are useful in the studio because they are insoluble, nonreactive, and inert. Pigment can only adhere to paper or canvas when it's mixed in some type of gluey binder.

That binder can be water-soluble, as in the case of watercolor, but the pigment itself remains a fine powdery dust.

In contrast to Earth's pigment particles, which are formed over millennia, the molecules of botanical dyes do not live in the soil. Instead, they are made within the body of the plant. They flow in aqueous solution through the plant and need live only as long as their host. Plants have the capacity to produce multiple biochromes that vary in hue and proportion depending on soil content, rainfall, and time of harvest. Altered by sunlight, acids and alkalis, enzymes, and minerals, they are soluble, reactive, and expressive.

But why transform fragile botanical dyes into pigments at all? Simply put: It allows us artists to extend the spectrum of the Earth's natural color palette into various shades of red, purple, blue, and yellow that aren't otherwise achievable. Each plant contains multiple constituents, and many hues can be drawn from a single source.

Most of the botanical pigment recipes in this book are a variety called "lakes," which are made by attaching soluble dye to an insoluble mineral carrier. This gives the colors substance, prevents them from dissolving, bolsters them against ultraviolet decay, and stabilizes their fragility. The carrier is produced from alum, an ingredient familiar to natural dyers. Along with pigments like Egyptian blue, vermillion, and smalt, lakes are early man-made colors—dating back to medieval Europe or earlier—and they demonstrate an inventive manipulation of natural ingredients. Much of my research relies on early dyers' and artists' manuals written at the crux of preindustrial craftsmanship and proto-chemistry. The ingenuity of our artisan forebears has directly inspired many of the recipes in this book.

When I first began researching how to make lake pigments, I thought the process required a centrifuge, a vacuum filter, and an advanced degree in chemistry. Fortunately, historical recipes transported me to a medieval artist's studio, where a well-worn piece of flannel was the filter paper, and a seashell served as a paint pan. Simple age-old tools, along with a craftsperson's attentiveness to materiality, are all we really need to make lake pigments. But the process takes several days—primarily passive time consisting of filtering and drying. Throughout the book, I share my favorite tools, as well as some modern conveniences to help you enjoy a smooth process.

In Chapter 3, the Master Lake Pigment recipe (page 63) serves as your primary springboard to artistic exploration, supported by chemical explanations that are intended to empower you in your efforts.

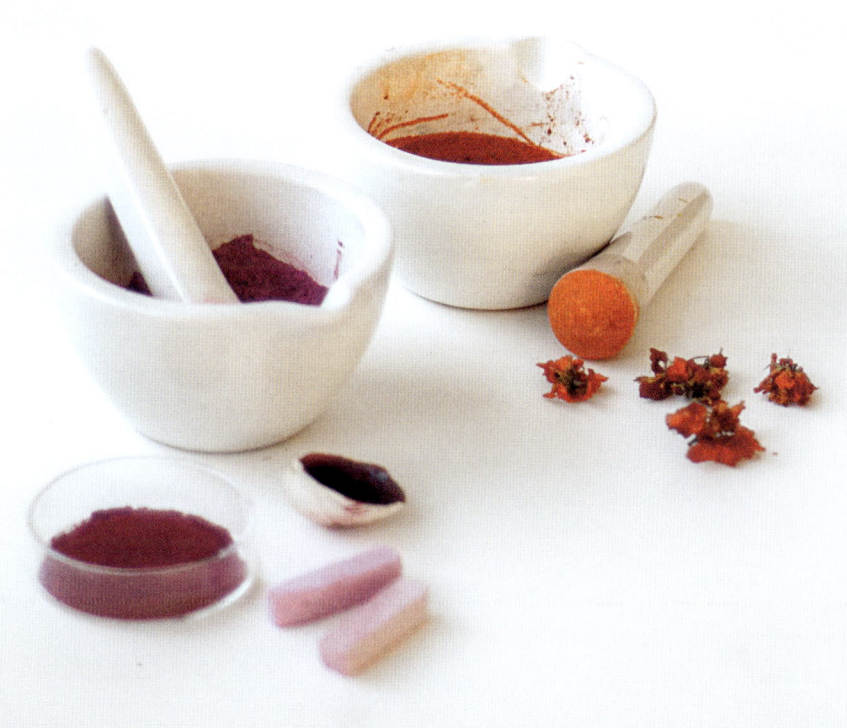

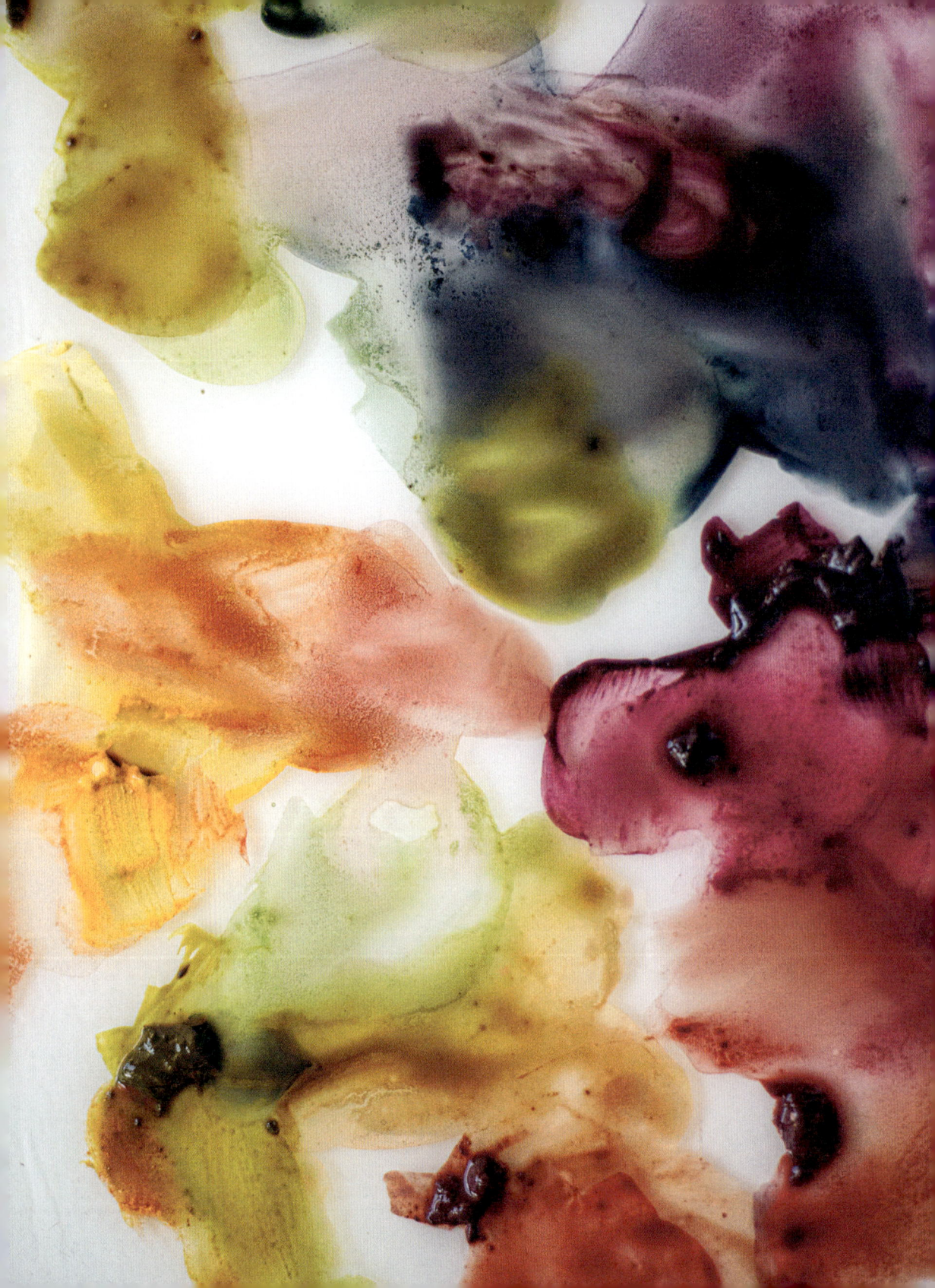

The recipe can be applied to any plant listed in the Dyestuff Ratios Chart (page 72) to create a wide spectrum of color. Like a chef riffing in the kitchen, tinkering with each element of the recipe allows for even greater nuance. To that end, the Lake Pigment Variations (page 79) demonstrate alternative methods of dye extraction and highlight specific dye sources suited to each process. Chapter 4, Other Plant Pigments, introduces methods beyond laking for botanical colors. Ingredient substitutions are offered in Chapter 2, the Studio Larder (page 26), further broadening the working characteristics and regionality of our lake pigments. There are many uses for botanical pigments in the studio. Like other pigments, they can be used to color tempera, oil paint, watercolor, plaster, crayons, encaustic, ink, varnish, or any pigment-compatible media. Manual pigment processing and safety guidelines are addressed in Chapter 5, Artists' Materials (page 146). These recipes for ink, watercolor, and pastels are a practical introduction to the wide realm of handmade artists' media.

In our current ecological reality, plant networks are stressed by climate extremes and introduced species, some of which are legally designated noxious weeds or invasive. Learning to identify and responsibly utilize overabundant plants offers a way to find beauty even in humble weeds. Keep in mind: Ethical foraging is just one way to source dye plants; whether your dyes come from your garden, a vendor, or an abandoned lot, bringing intentionality to your material choices is empowering. I discuss this more in Chapter 1, Botanical Sources.

Art speaks a visual language of color, texture, form, and surface. Botanical pigments do all this, and they offer far more than just visual beauty and appeal. Nurturing, learning from, and fostering kinship with plants links us to a longstanding heritage. It reclaims a connection to nature, and rekindles wisdom, that has largely been severed in the modern world. Each botanical pigment is not just a visual representation of nature, it is also a collaborative manifestation of nature. These colors are rich with history, entangled in ecological networks, and alive with possibility.

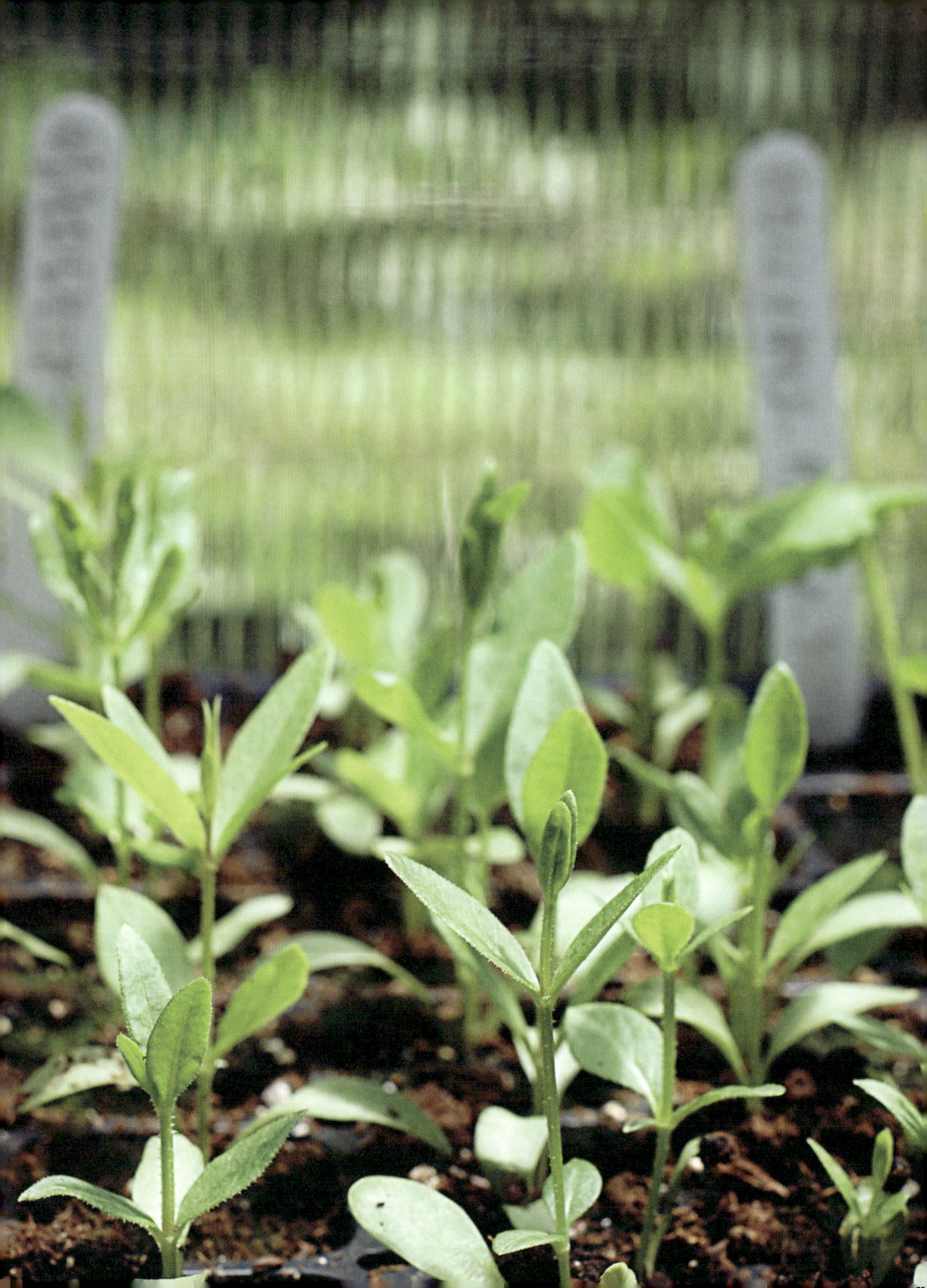

1

BOTANICAL SOURCES

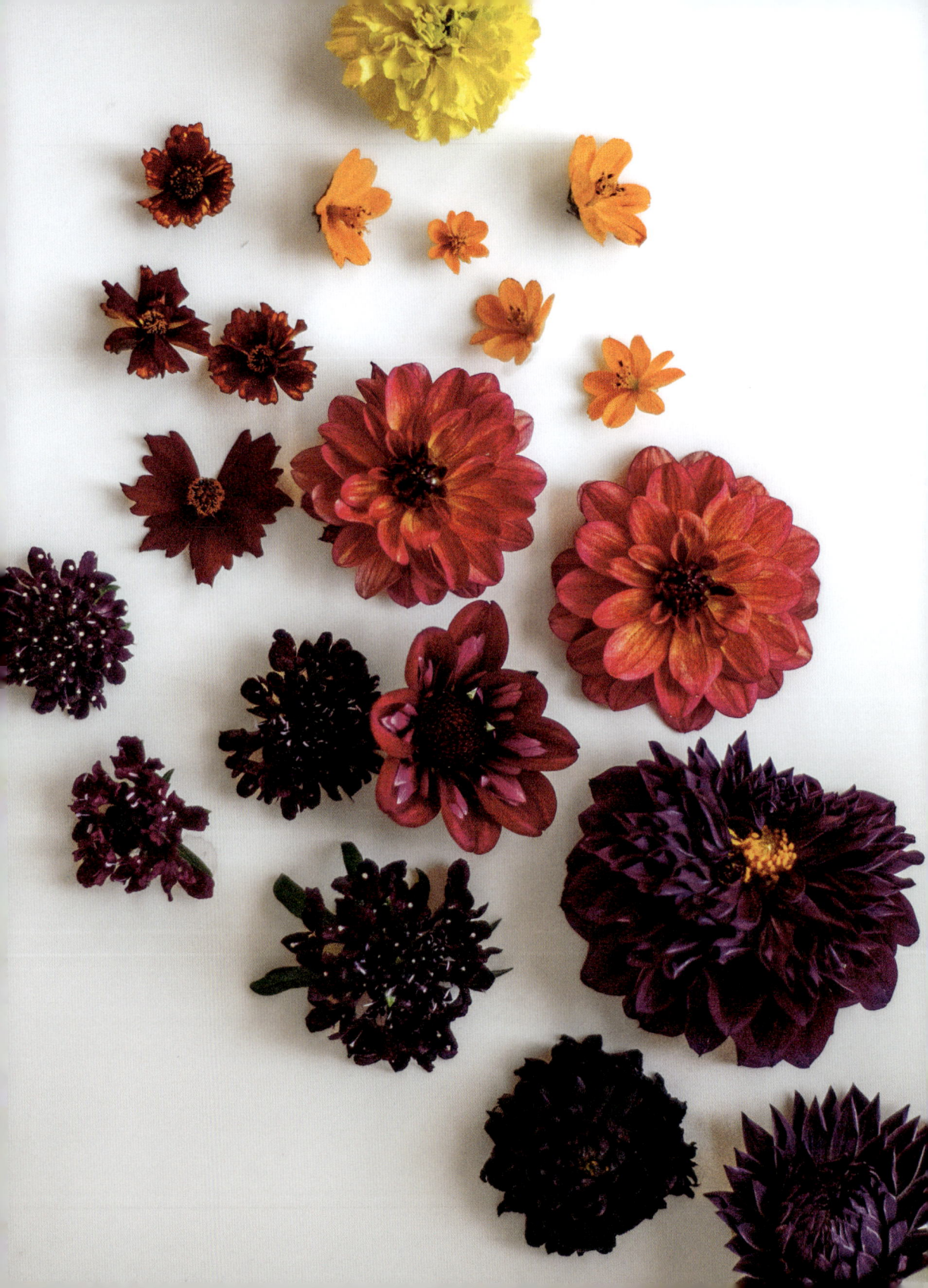

All botanical pigments start with plants. In this chapter, I share an overview of some undemanding dye plants that are ideal for beginners to cultivate in the garden. I also include helpful guidelines for foraging plentiful species. Foraging, drying, and storing plant material follows a seasonal rhythm of being outdoors during the growing months and indoors, next to steaming dyepots, in the fallow season. Following the thread of abundance in what and when we harvest influences the colors we concoct, reflecting the seasonal landscape local to each of us.

Cultivating Color

There are three ways to source dye plants: purchasing, cultivating, and foraging. There are many knowledgeable dye vendors who offer a wide variety of plants. When working with a vendor, make sure to seek out a specialist who can provide information on the specific plant species, places of origin, and best practices for working with them. Because dye plants are a natural resource and commodity, their provenance and production should be considered.

Many dye plants can be grown at home or in a community garden as well as foraged locally. These options foster a connection to place, demanding care and observation.

Whether purchased, cultivated, or foraged, any plants used for making lake pigments must contain mordant dyes, which bond to the mordant alum (page 37) forming the pigments' core. As you flip through seed catalogs and familiarize yourself with local species, compare those plants to the varieties listed in this book and in other dyeing resources.

In the following pages, I've included a sampling of plants to get you started. These annuals, biennials, and perennials are perfect if you're just beginning your journey into making lake pigments. Each yields a variety of different colors, some of which I demonstrate later in the book. Marigolds, for example, are a colorful and floriferous option for beginners; they're very forgiving of neglect yet continue blooming all summer long. Before anything else, look up your local plant hardiness zone. This number corresponds to the average extreme temperatures in your region and is used to determine which varieties will thrive under your growing conditions. When selecting seeds, make certain your hardiness zone falls within the range suggested for each plant variety. It's helpful to source seeds cultivated in a climate like your own, be it from a regional seed vendor or a seed-savers' swap. Follow the guidelines for sowing, sun exposure, and water needs suggested by the seed supplier. This will give your plants a good start so they will grow happily in your local conditions.

It's also important to familiarize yourself with any regionally designated invasive species. Take care not to cultivate these, and remember that not every plant listed in this book is appropriate for every garden.

Easy annuals

These plants are grown from seed in the spring and complete their entire lifecycle in one growing season. Leave some flower heads to mature and set seed in autumn; you can save the seeds in a cool, dry place to perpetuate your dye supply the following year.

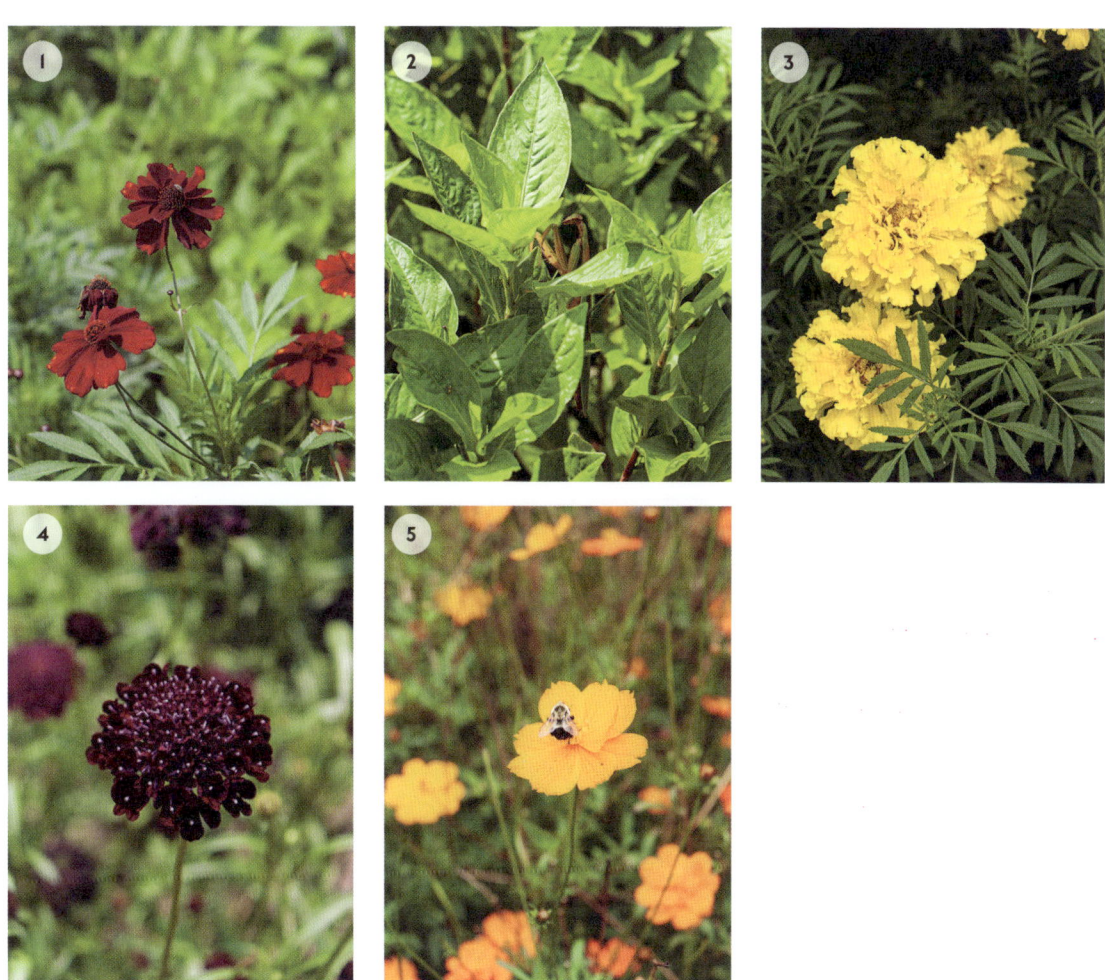

1 Dyer's Coreopsis blossoms or plant tops (*Coreopsis tinctoria*)
2 Japanese Indigo leaves (*Persicaria tinctoria*)
3 Marigold blossoms or plant tops (*Tagetes* spp.)
4 Pincushion flower, dark purple blossoms (*Scabiosa* spp.)
5 Sulfur Cosmos blossoms (*Cosmos sulphureus*)

Biennials

Taking two years to mature for harvest, biennials require a little more patience than annuals. After gathering energy over an entire growing season, they burst into flower in their second year, often with a considerable leg up on annuals.

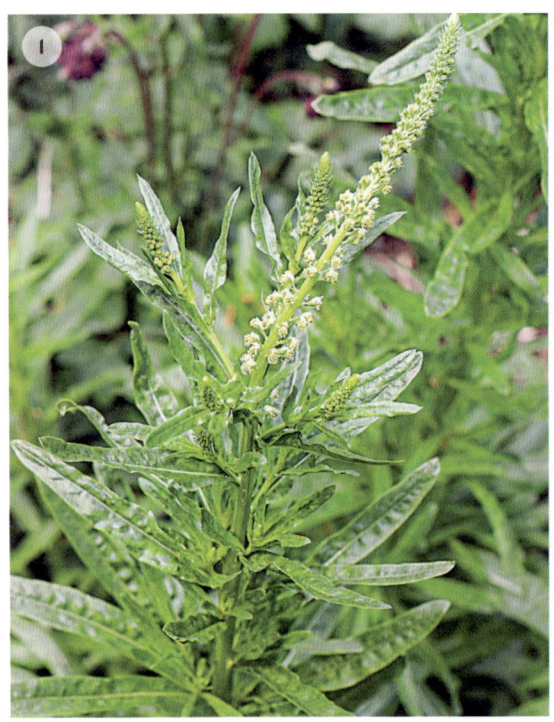
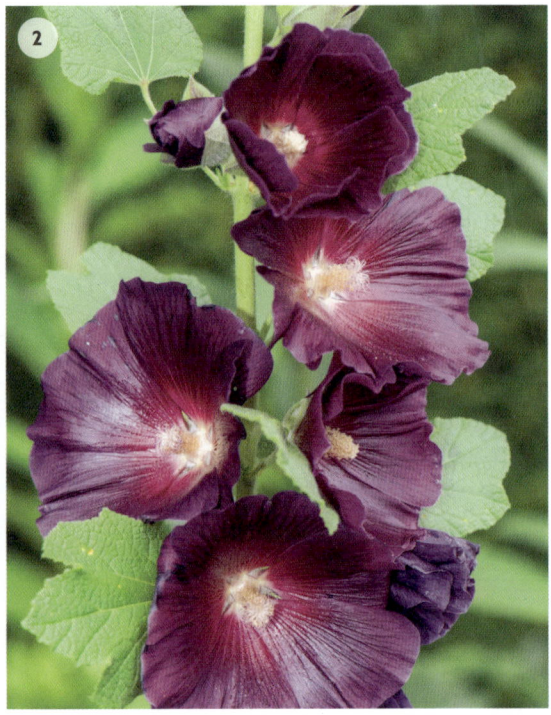

1 Weld flowering spike (*Reseda luteola*)
2 Hollyhock blossoms, dark purple (*Alcea* spp.)

OTHER PERENNIALS TO TRY

Once you've had some success growing dye plants, you may be emboldened to expand your horizons. Why not try growing turmeric rhizomes (*Curcuma longa*), butterfly pea flower (*Clitoria ternatea*), dyer's greenweed (*Genista tinctoria*), hardy hibiscus (*Hibiscus* spp.) and rhubarb root (*Rheum* spp.)?

Perennials

It might take two to three years to reap a harvest from seed-grown perennials, but they return year after year with a bounty of color. Get a head start—and make sure your variety is true to type—by purchasing an established plant or bartering for divisions from gardening friends.

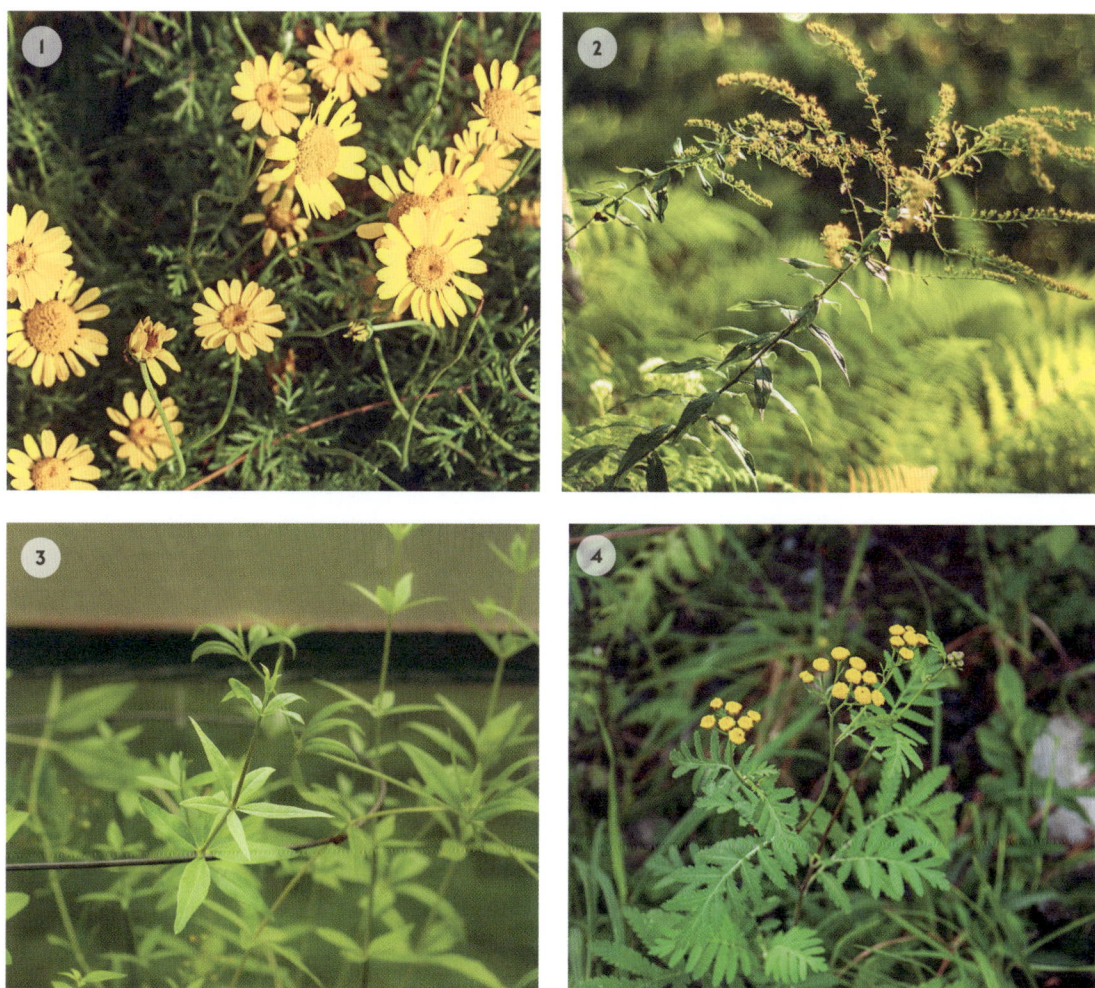

1. Dyer's Chamomile blossoms or plant tops (*Cota tinctoria*)
2. Goldenrod blossoms or plant tops (*Solidago* spp.)
3. Madder root (*Rubia tinctorum*)
4. Tansy blossoms or plant tops *(Tanacetum vulgare)*

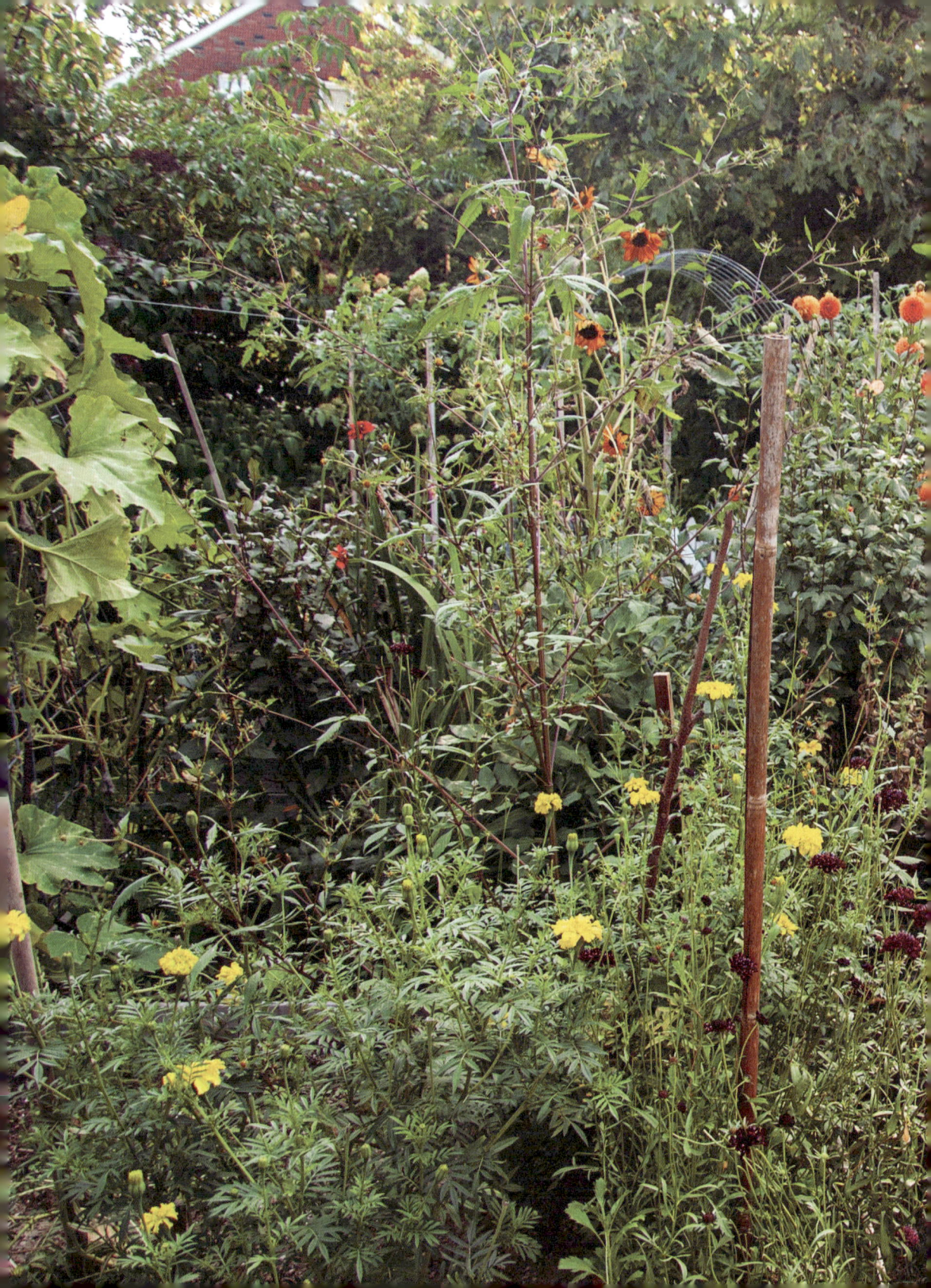

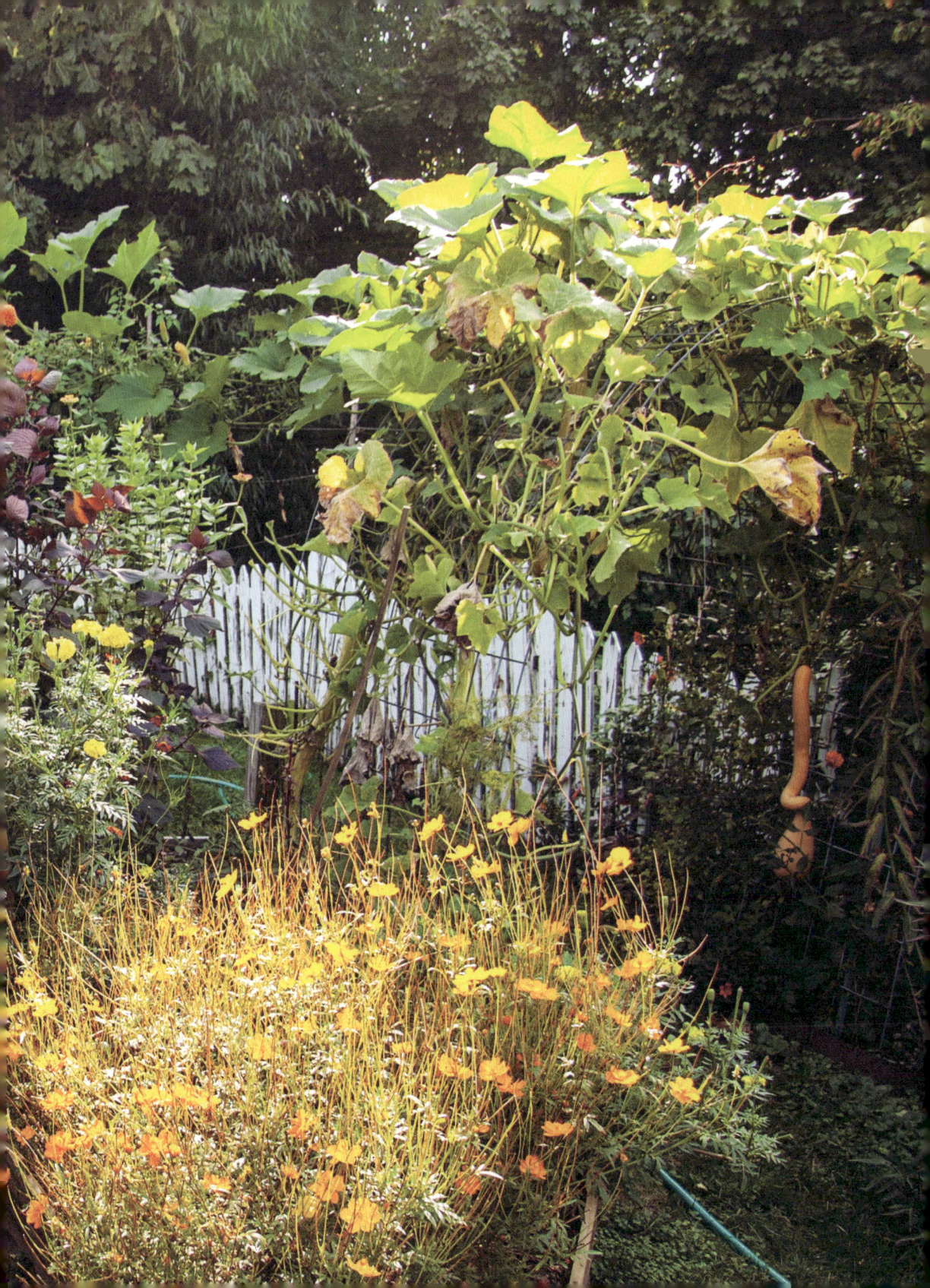

Foraging Ethics

Wild plants provide a place-based palette slightly different from their cultivated counterparts. Foraging comes with the ethical responsibility to enjoy the bounty of nature without harming ecological networks. First, consider where you can access a wild harvest and how to do so safely.

Getting started with foraging can be daunting. The best way to learn the lay of the land and practice plant identification is alongside other knowledgeable plant-lovers. Consider joining a local foraging group or mycological society, volunteering for community cleanup days in green spaces and waterways, or joining your regional invasive species management group. There also are many helpful plant identification apps and field guides available.

DISPOSING OF INVASIVE SPECIES MATERIAL

Boiling invasive species plant material for dye extraction should destroy its regenerative potential. If you have any doubts or surplus dyestuff, it should be buried 2m (6ft) deep or disposed of in the garbage. Check with your local authorities or invasive species management group for guidelines on specific plants.

FORAGING DOS & DON'TS

- Do research the foraging regulations in your local public lands and parks.

- Do identify plants using multiple characteristics and know the common lookalikes to avoid.

- Do familiarize yourself with the endangered and threatened plants in your region, and never harvest these.

- Do get to know the invasive species in your region.

- Do protect yourself from the elements, such as ticks, poison ivy, sunburn, and so on.

- Do ask permission before harvesting a plant.

- Do observe the health and stature of the plant or plant colony. Take note of any other creatures that might be using the plant for food or habitat.

- Do not forage on private property, except with explicit prior permission.

- Do not gather plants along roadsides or anywhere with evidence of herbicide use.

- Do not harvest more than one-third of a plant or group of plants. Take only what you can respectfully utilize or responsibly dispose of.

Invasive Species

Invasive species are a legally designated group of introduced plants (and other creatures) that displace native species, suppress biodiversity, and disrupt the co-evolved relationships of an ecosystem. Regional regulations define which species are labeled invasive, as well as the measures that can be used to control their spread. Where you live, these plants might be variously termed noxious weeds, non-native species, or invasive alien species. A hierarchy of alert levels can further divide plants into subgroups.

Working with invasive species demands we acknowledge that the commodification of natural resources, intentional plant introductions, and climate change are causing species to migrate. These plants aren't inherently evil; they are resilient and unfettered by the bonds of ecosystem relationships. Because of their overabundance, invasive species are some of the best plants to forage for color, if done responsibly. Utilizing overabundant species for botanical pigments is one step toward encouraging biodiversity, developing care for plants, and finding the beauty in weeds.

FORAGING INVASIVE SPECIES

- Do look up the invasive species in your area to see if any of them contain mordant dyes.

- Do learn the reproductive pathways of these plants, and take care not to inadvertently spread material to new areas.

- Do not intentionally transplant or propagate invasive species.

- Do harvest what you can reasonably use or responsibly dispose of.

- Do feel free to harvest an entire plant, or even a whole colony of plants, but have a succession plan to fill that gap with vigorous native plants.

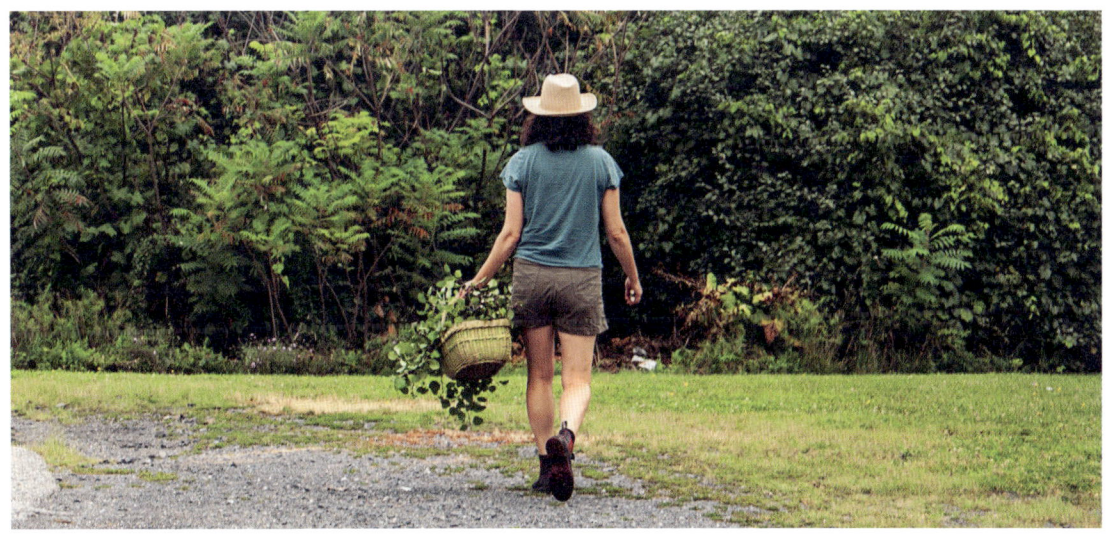

Harvesting & Storing Plants

Dyestuff can be used fresh or dried for storage. Choose a well-ventilated, dry space out of direct sunlight. Spread the blossoms or chopped-up dyestuff on sheets of newspaper, screens, or a multilevel herb-drying rack. Turn the plant material every day or two until completely dry. Store dye plants indefinitely out of sunlight in paper bags, burlap sacks, or similar breathable material.

Many annual dye plants are grown for their blossoms, which can be gathered all summer long. A rolling harvest every few days prevents the plant from setting seed and encourages reblooming until frost. Individual blossoms can be snapped off as they just start to fade. (I like to give the bees and pollinating insects the chance to enjoy them first.) To keep an orderly look in the garden, you might like to trace the stem down to the next pair of leaves and snip it off just above the leaves, from which point the plant may produce more flowering stems. Blossoms may be left whole, with stems removed.

Some flowering dye plants also contain rich color throughout the plant top: the leaves and stem above ground. It can be a different shade or less vibrant than the blossoms alone, so you might prefer to separate the two. Leaves and stems should be harvested well into the season when they are growing robustly, but before the plant's energy is diverted into producing seeds or dwindles into winter dormancy. Use garden shears to chop fresh leaves and stems into pieces, roughly 1½ to 2½cm (½ to 1in) long.

Accessing dyes contained in heartwood, bark, or roots requires the pruning or death of a plant. Perhaps you manage a woodlot, maintain a fruit orchard, or are mitigating invasive species. In this case, you might find yourself with a wonderful glut of dye material. Heartwood should be rasped or shaved into small pieces. Roots should be washed in cold water to remove all dirt, and then chopped into segments while still fresh.

If bark is your aim, you'll want the inner strip of spongy bark, or the cambium layer, pulled from the largest possible limbs or trunk. This will wound or kill a living tree, so strip bark only from pruned limbs or felled trunks. Working from recently cut wood, remove and discard as much of the craggy outer bark as possible. Shred or chop the inner bark into pieces, roughly 1½ to 2½cm (½ to 1in) long.

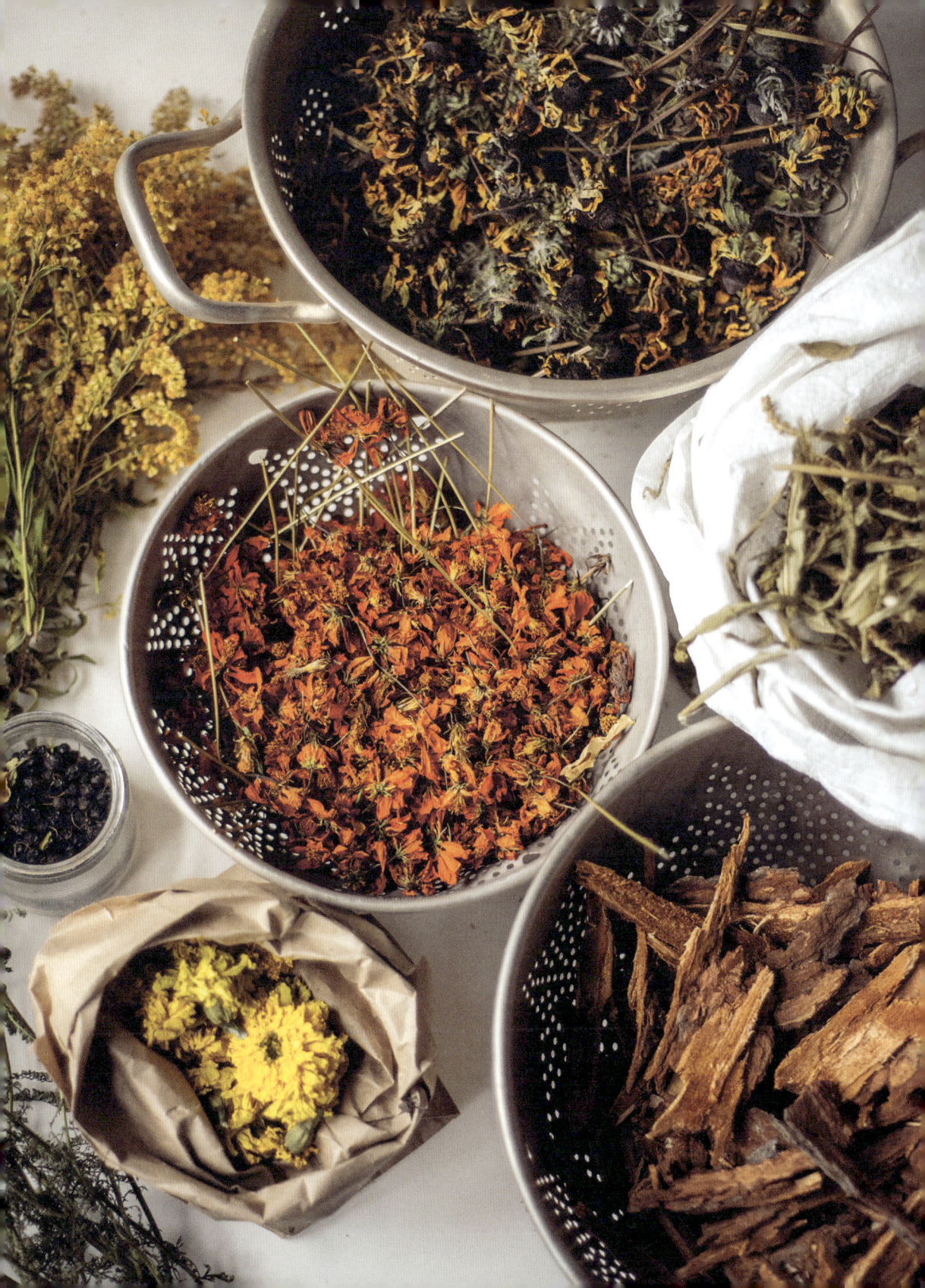

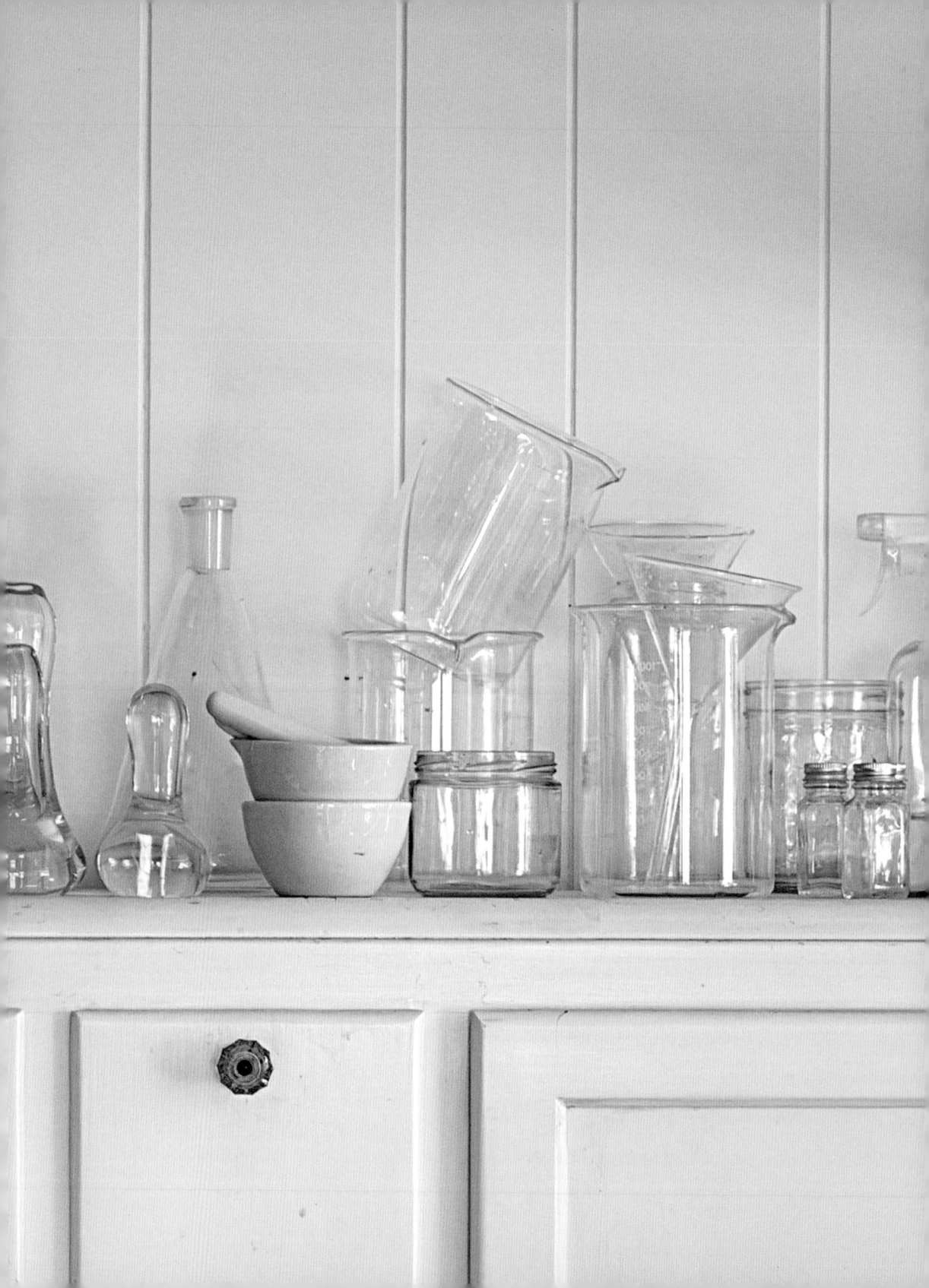

2

THE STUDIO
LARDER

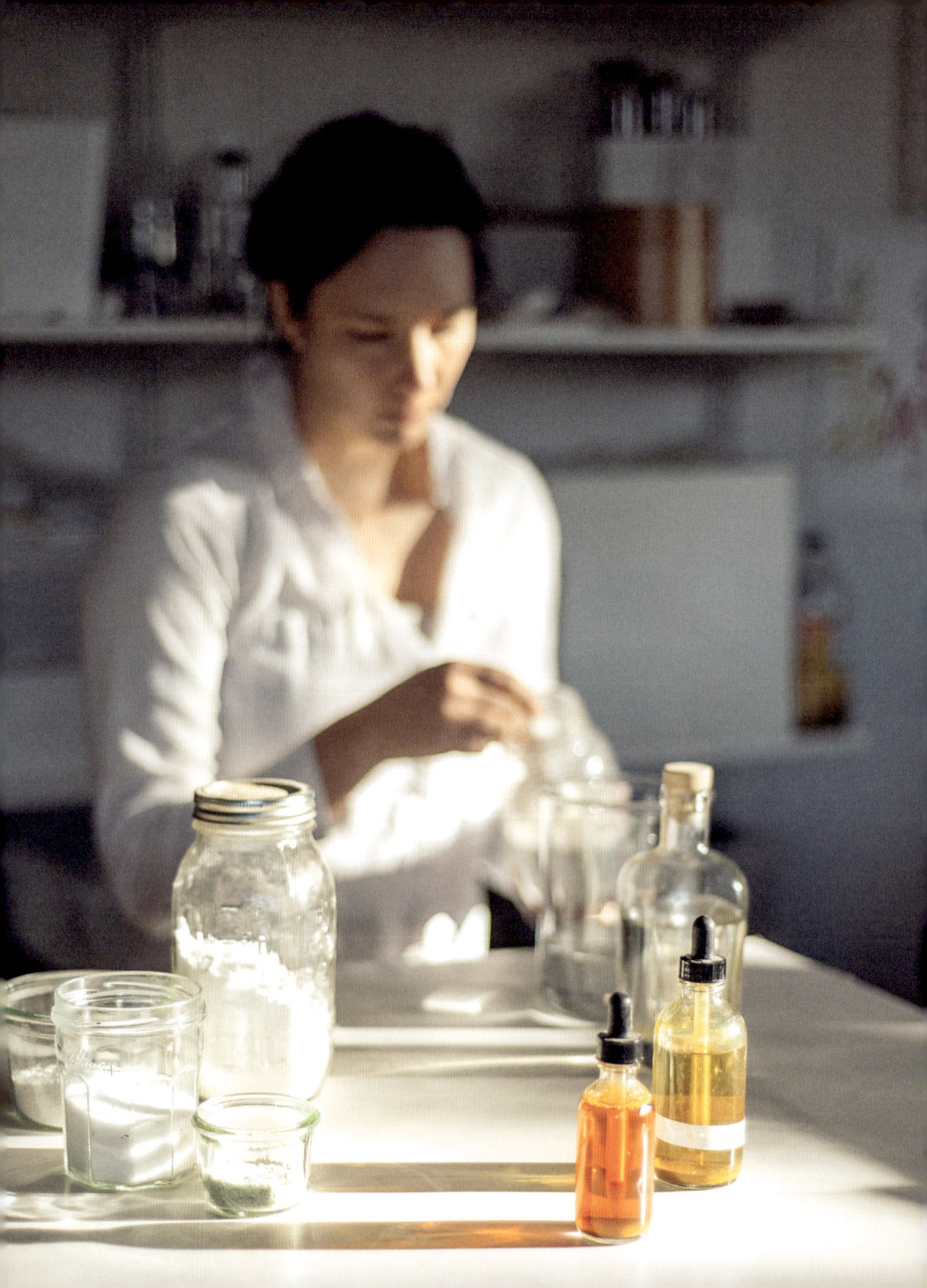

A well-provisioned studio cupboard allows us to whip up a feast for the eyes. Beyond a cache of dyestuff, all that's needed to concoct plant pigments are some tools, a water source, and a few powders. These mordants and bases are the chemical means by which botanical dyes are transmuted into pigments. In this chapter, I provide an essential equipment list and an introduction to the ingredients—along with a few thrifty substitutions.

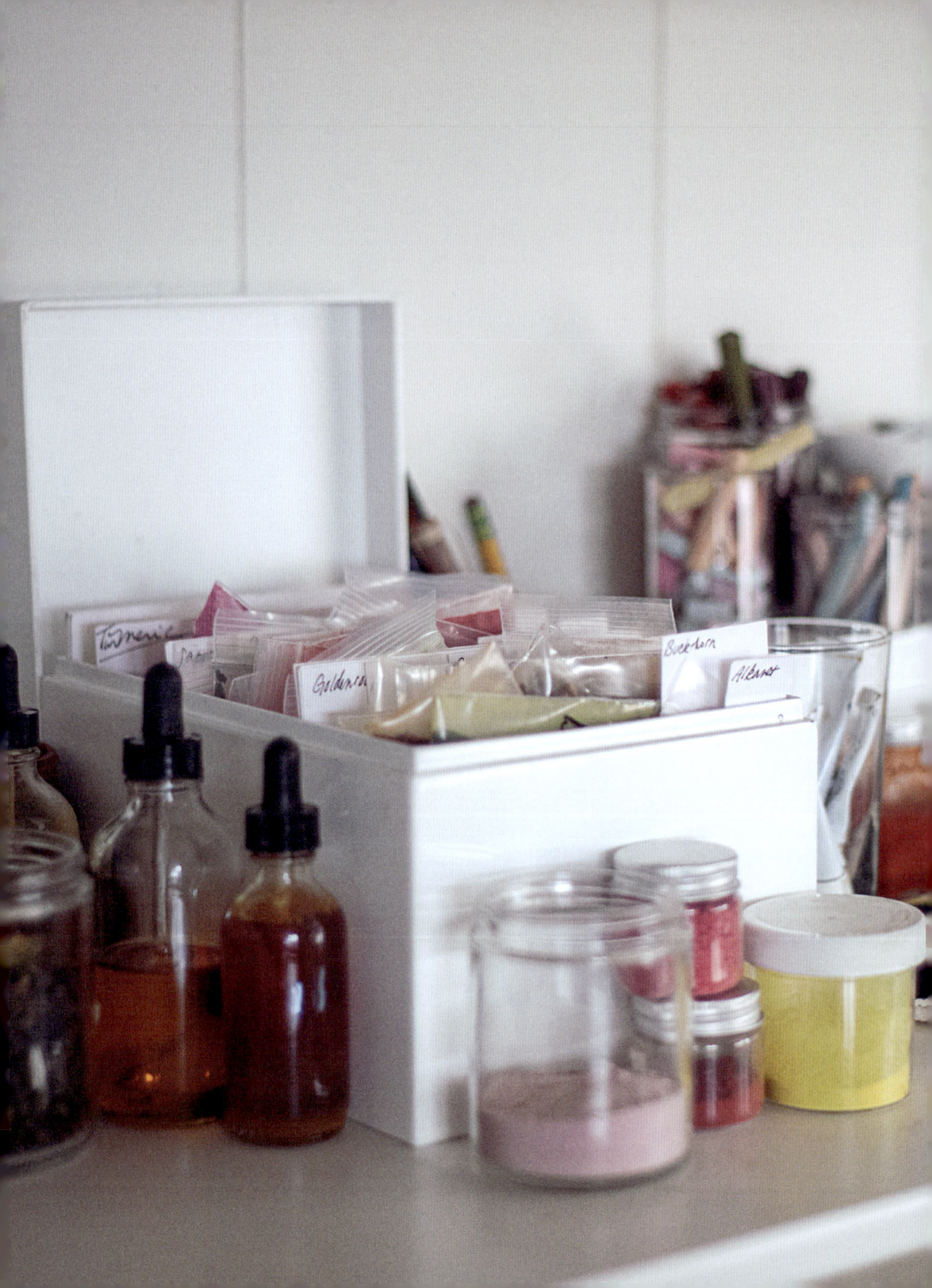

Tools & Materials

All tools must be dedicated to artmaking and not borrowed from the kitchen. Dye colors are exceedingly sensitive to iron, so make sure that all cookware and tools are completely rust-free. You will need the following items:

1. Metric digital scale accurate to 0.1 grams

2. Stainless-steel stirring spoon

3. Whisk

4. Stainless-steel tongs

5. Medium stainless-steel sieve or funnel

6. Stainless-steel pot, 1 liter/1 quart

7. Spatula or palette knife

8. Thermometer

9. Spray bottle of water

10. Heatproof wide-mouth jars or beakers, 1 liter/1 quart, minimum 2

11. Paper coffee filters, 20-cm (8-in), round

12. Gloves, such as nitrile gloves

13. Synthetic mesh filter, such as dashi sachets, cold brew filters, or nut milk bags

14. pH indicator strips or pH meter

15. Small dish or beaker

16. Lidded storage jars

17, 18. Tape and marker for labeling

· N-95 face mask or respirator (not shown)

· Safety goggles (not shown)

KEEPING NOTES

I highly recommend keeping notes on your experiments in a dedicated sketchbook or notebook. I number each pigment with a masking-tape label, which follows the pigment from pot to jar to strainer. I pair that label with the weight of each ingredient, extraction details, a running tally of what steps I've accomplished, and any novel observations I have. Swatches of both the dye and pigment can go here too!

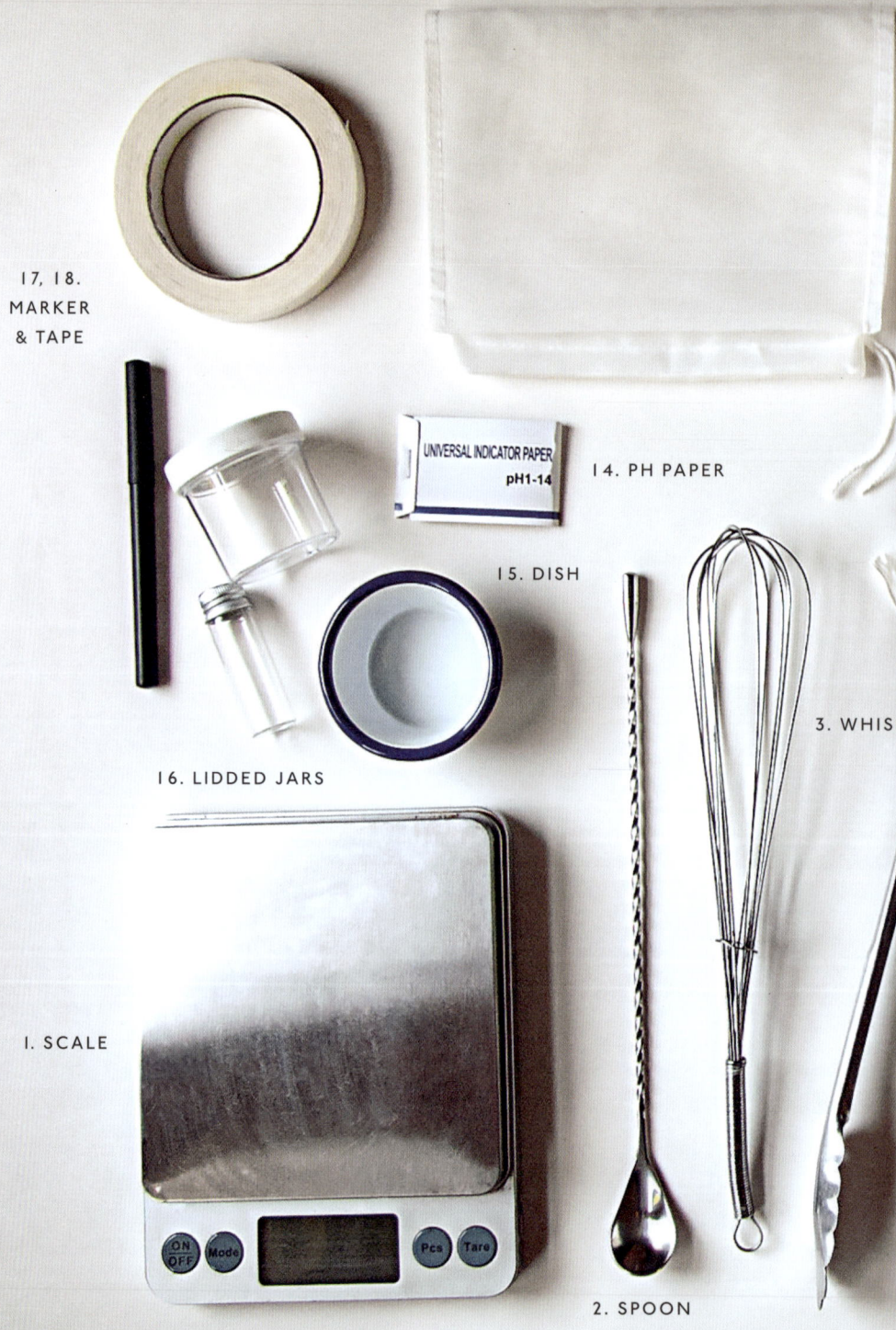

13. MESH FILTER

17, 18.
MARKER
& TAPE

UNIVERSAL INDICATOR PAPER
pH1-14

14. PH PAPER

15. DISH

3. WHISK

16. LIDDED JARS

1. SCALE

2. SPOON

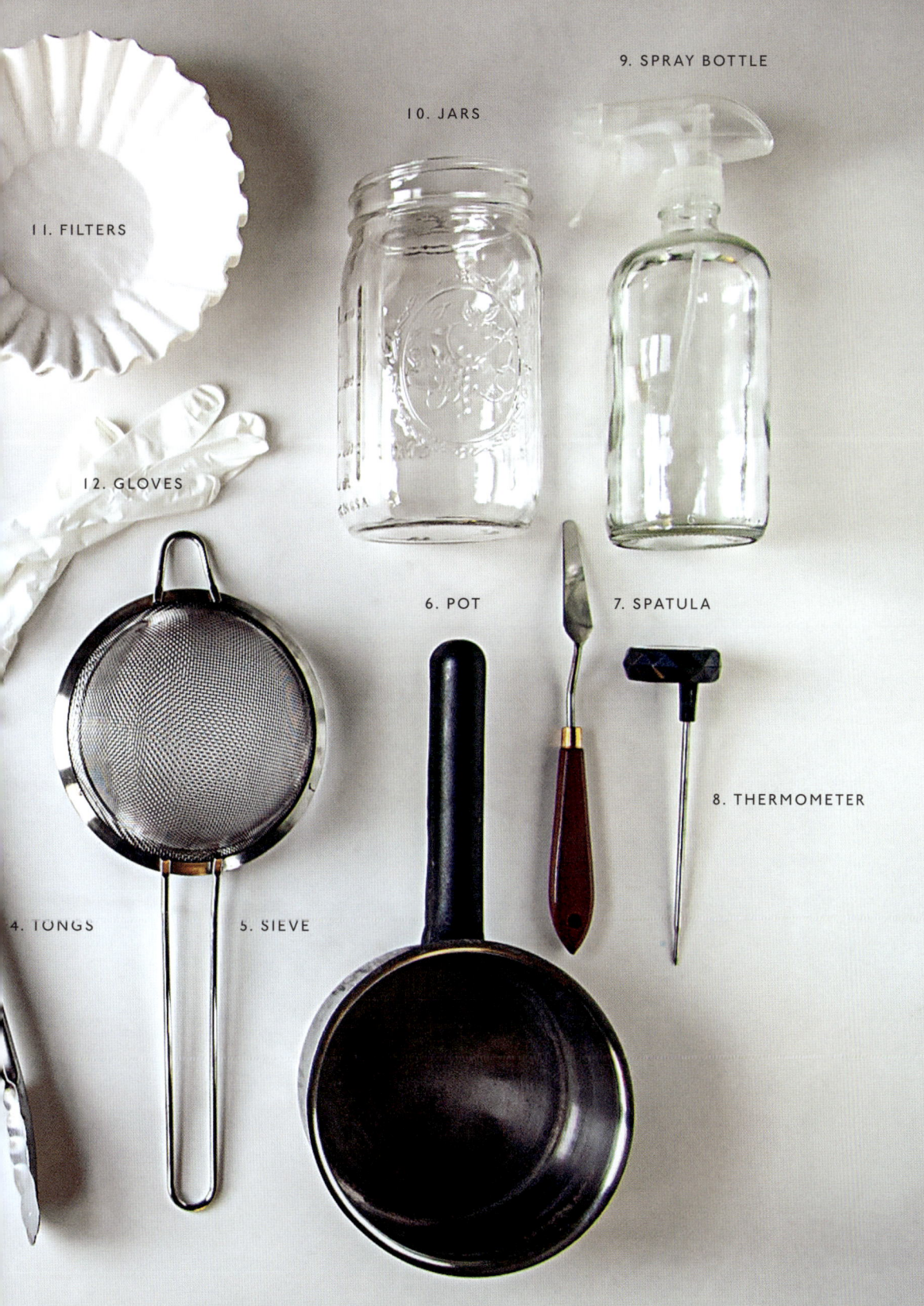

9. SPRAY BOTTLE

10. JARS

11. FILTERS

12. GLOVES

6. POT

7. SPATULA

8. THERMOMETER

4. TONGS

5. SIEVE

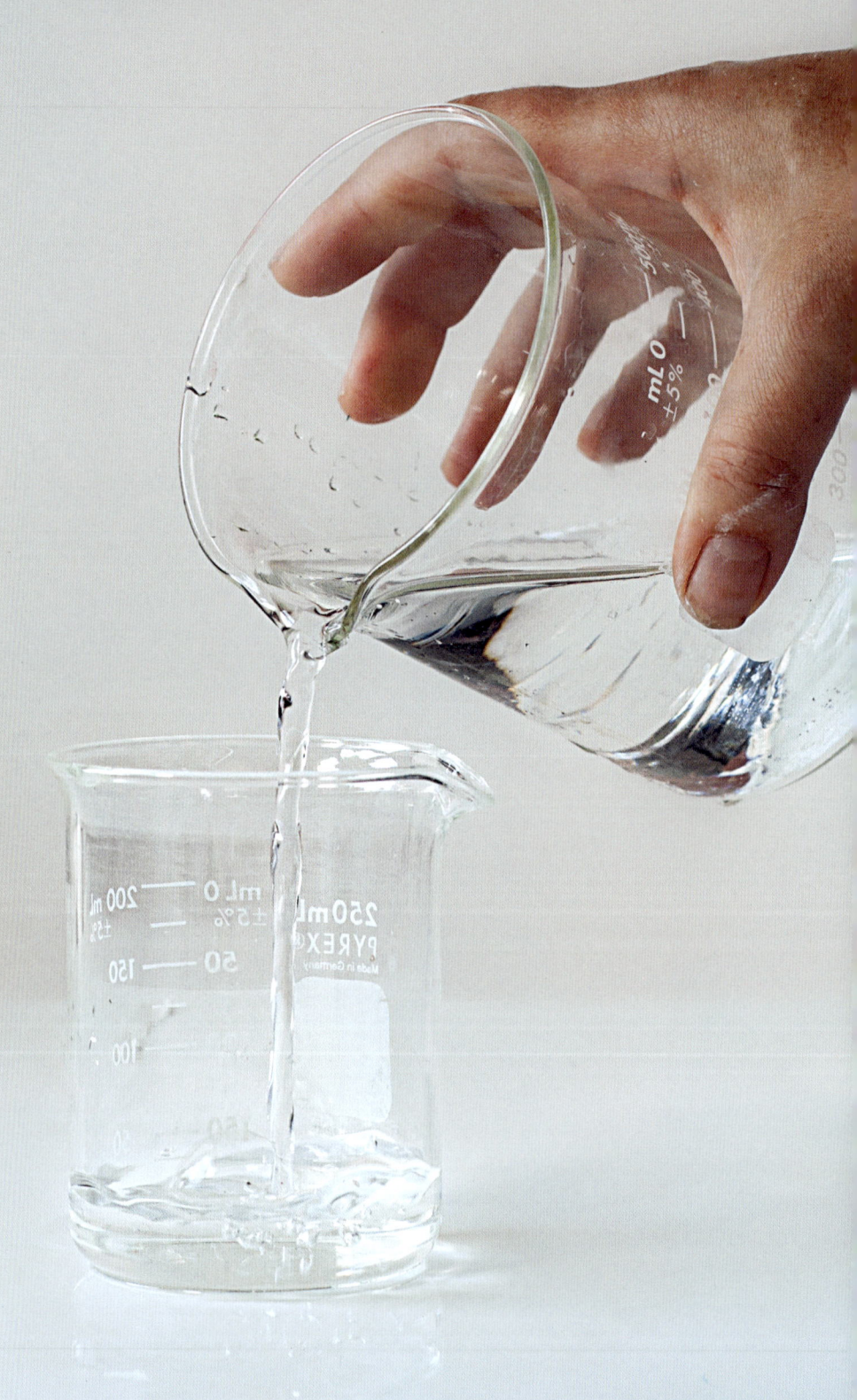

Water

A pure water source is essential to ensuring the clarity of color and the success of chemical interactions. Water tinged with iron neutralizes dye colors, sapping their vibrancy. Soft water can leach iron and other metals from soil, pipes, and pots. The calcium present in hard water can affect pH and alter the shade of natural colorants. If you have any concerns about the purity of your water source, avoid disappointment by working with captured rainwater or distilled water. Municipal tap water, so long as it is neither very hard nor iron-laden, is sufficient for most pigment-making processes (occasionally, I specify otherwise). Washing finished pigments in distilled water is a good practice to guarantee their purity. Fermentation requires a water source without added chlorine or chloramine, which deliberately inhibit microbes.

HEALTH & SAFETY

All studio ingredients, whether natural or store-bought, should be used with common-sense precautions and a healthy respect. Keep the contents of your studio larder clearly labeled in a place secure from curious children and pets. Finished lakes and other botanical pigments are chemically quite safe; some are used commercially to color food and cosmetics. It's important, however, to protect our respiratory systems from these fine powders. Store all dry ingredients in airtight containers, and wear respiratory protection when grinding pigments and handling fine powders including chalk and slaked lime. When mixing dry and wet ingredients, avoid sending up a cloud of dust by adding powder to liquid rather than the reverse. Wear gloves when handling mordants and bases.

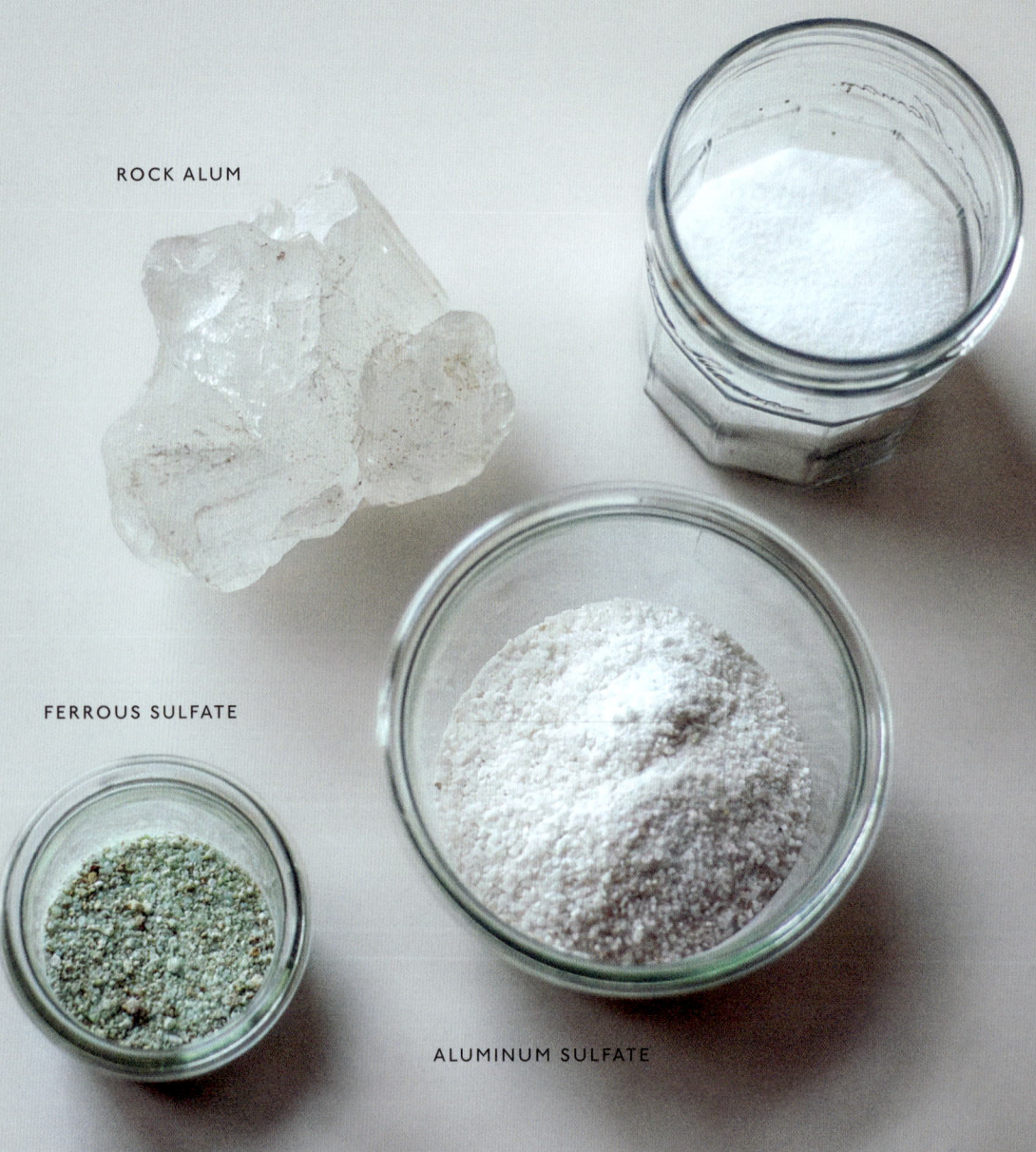

POTASSIUM ALUMINUM SULFATE

ROCK ALUM

FERROUS SULFATE

ALUMINUM SULFATE

Mordants

Mordants are metallic salts with the chemical tendency to bond with other molecules, notably natural dyes and fibers. The term "mordant" is descended from Latin for "to bite," as these salts permit dyes to latch permanently onto a fiber carrier. Alum and iron are two mordants used in pigment making.

Alum

The mordant commonly called "potash alum," or simply, "alum," is potassium aluminum sulfate $[KAl(SO_4)_2 \cdot 12H_2O]$, a hydrated double salt of aluminum that has long been used in dyeing, papermaking, and leather tanning. Aluminum sulfate $[Al_2(SO_4)_3 \cdot 16H_2O]$ is a relative newcomer to dyeing, synthesized without the somewhat extraneous potassium.

Both potassium aluminum sulfate and aluminum sulfate contain similar amounts of aluminum and can be used interchangeably for lake pigments in equal measure. As in natural dyeing, the alum used to make pigments likewise serves to stabilize dye colorants against dissolution in water and fading in sunlight.

Always source alum from a reputable dye vendor, who will prioritize its purity and the vibrancy of your pigments. Avoid aluminum acetate mordants and pickling alum, which are not used for pigment purposes.

Iron

Iron is both a mordant, like alum, and a color modifier. It saddens (darkens) natural dyes by shifting their hues to deeper, more neutral shades. This striking and irreversible color shift is the reason behind dyers' vigilance for warding off rusty tools and iron-tinged water. If you seek a palette of subtle and dusky neutrals, iron is a wonderful ally!

FERROUS SULFATE SOLUTION

Soluble iron powder called "ferrous sulfate" is available from dye vendors, or you can concoct a homemade iron solution (opposite page). This 10-percent ferrous sulfate solution has a dependable potency. Take care not to store iron solution in a container with a metal lid; it may quickly rust through.

MATERIALS

· 10g ferrous sulfate (green vitriol or iron [II] sulfate)

· 100ml distilled water, steaming hot

METHOD

1. In a small jar with a plastic or bale wire lid, combine the ferrous sulfate and hot water. The iron salts will dissolve readily.

2. Cap, label, and store at room temperature indefinitely.

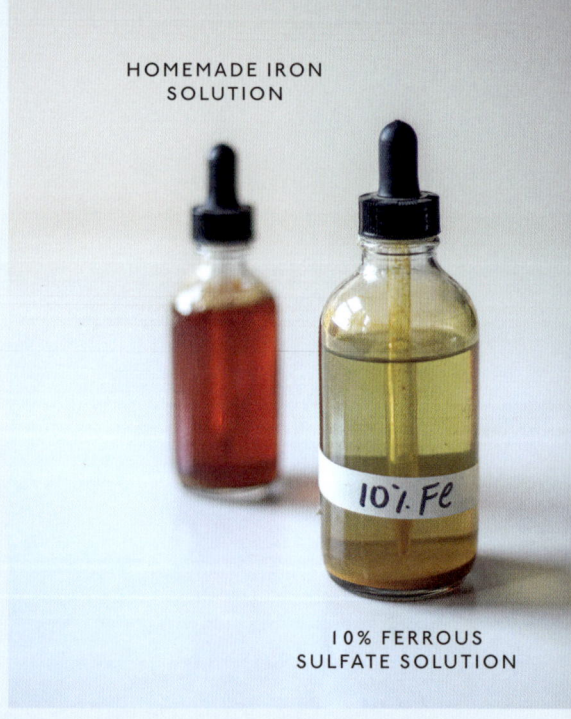

HOMEMADE IRON SOLUTION

10% FERROUS SULFATE SOLUTION

Over time, it is normal to see a thin film develop on the surface and a slight precipitation of the oxidized iron solution. To use, do not stir or disturb the sediment. Probe beneath the surface with an eyedropper to gather the liquid.

HOMEMADE IRON SOLUTION

A lifetime's supply of iron is contained in just a few scavenged rusty bolts and nails. Simply steep in vinegar, and you have a thrifty, homemade iron solution. The concentration varies, so apply judiciously to avoid a rusty cast to your pigments.

MATERIALS

· Rusty iron object, such as an old nail or a bolt

· White vinegar of any sort

· Pinch of table salt

METHOD

1. Find a dispensable glass jar that will accommodate your rusty object.

2. Fill the jar one-quarter full of white vinegar and one-quarter full of water. Add a big pinch of salt.

3. Place the rusty item in the liquid. Leave the jar uncovered in a safe place for one to two weeks. (Place on top of an old saucer—the liquid will stain if it overflows.) You'll notice tiny bubbles and the development of a crispy crust or rusty-colored foam on the surface of the liquid, as well as brownish sediment. These are good signs.

4. Filter the liquid into a jar with a plastic or bale wire lid. Cap, label, and store at room temperature indefinitely. Save your rusty object to use again!

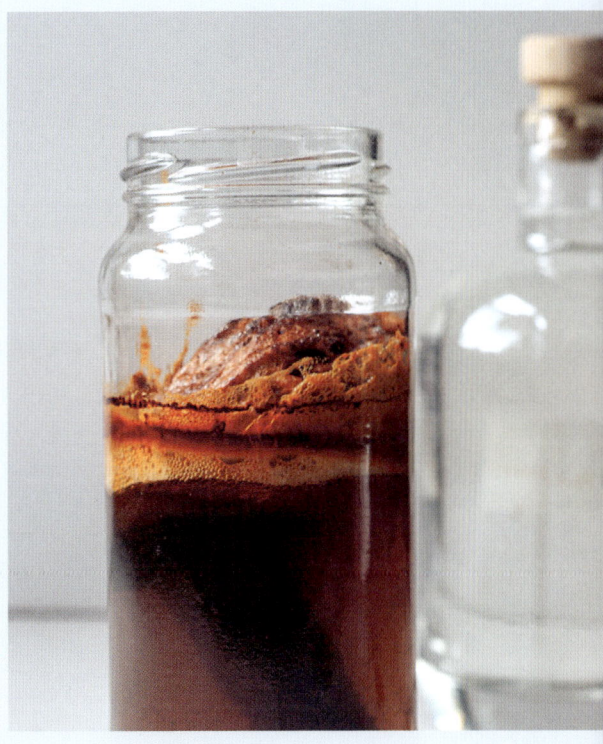

Be sure to use a deep-enough jar to allow for bubbling without overflow.

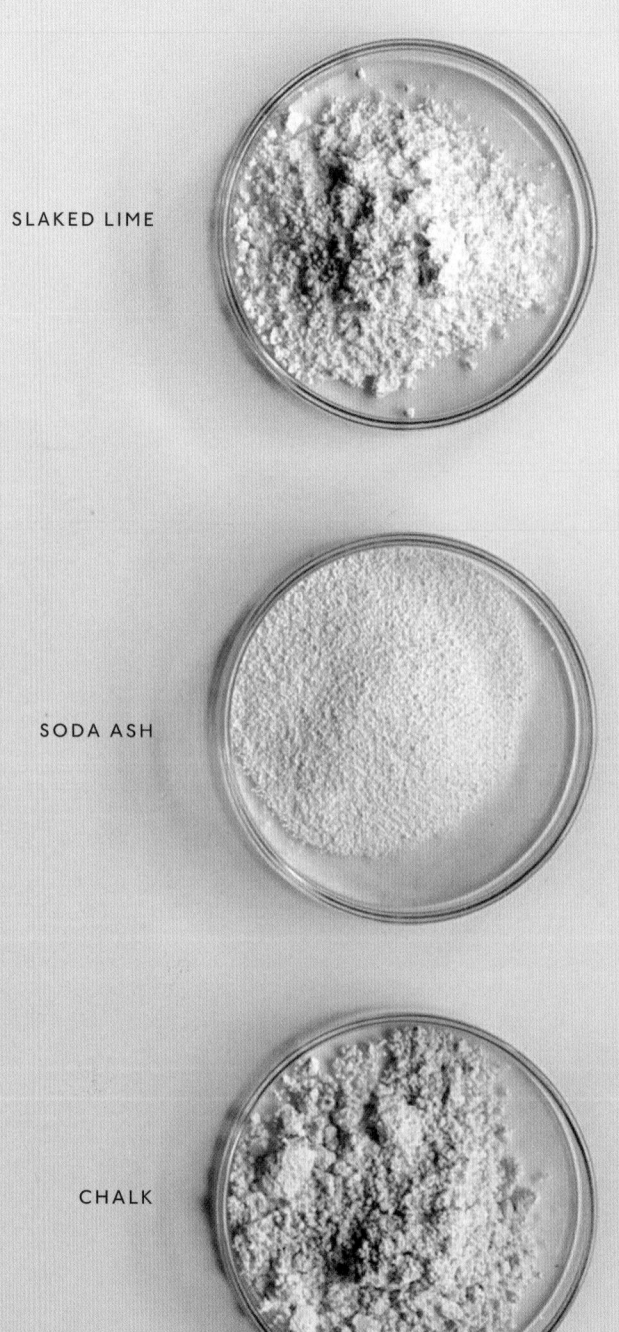

SLAKED LIME

SODA ASH

CHALK

Bases

Bases including soda ash, slaked lime, wood ash water, and chalk are used to manipulate pH, precipitating lake pigments from solution in the critical transformation from soluble to insoluble. The Master Lake Pigment Recipe (page 63) calls for soda ash to serve this essential function. Slaked lime, wood ash, and chalk are explored in Lake Pigment Variations (page 79) and Other Plant Pigments (page 124).

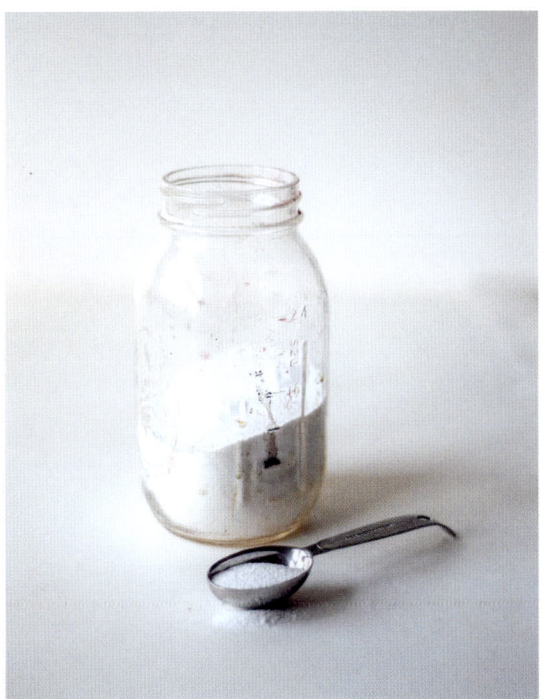

Soda Ash

Soda ash is the alkaline salt sodium carbonate, traditionally made by steeping the ashes of seaweed and other salty flora in water. It is readily available in dry, crystalline form.

In laking, soda ash is used at 50 percent the weight of alum, or enough to neutralize the dye/alum mixture. If the dye source or extraction method affects the pH, use pH indicator strips to pinpoint the correct amount. Because soda ash doesn't contribute body or opacity, the resulting lake pigment is translucent—perfect for watercolors and oil glazes.

If you don't have easy access to soda ash, you can substitute washing soda. This is a less potent hydrate of sodium carbonate, so use washing soda equal to 1.7 times the amount of soda ash specified. For example, 5g soda ash = 8.5g washing soda.

Slaked Lime

Calcium hydroxide is also known as calx, pickling lime, or slaked lime—this last term hinting at the compound's genesis. It is made by quenching highly reactive quicklime in water, poetically slaking its thirst. Calcium hydroxide is a caustic, finely powdered alkali that can be used for both precipitating lake pigments and flocculating indigo. Keep it in a safe spot inside a well-sealed container, and wear a mask and gloves when handling.

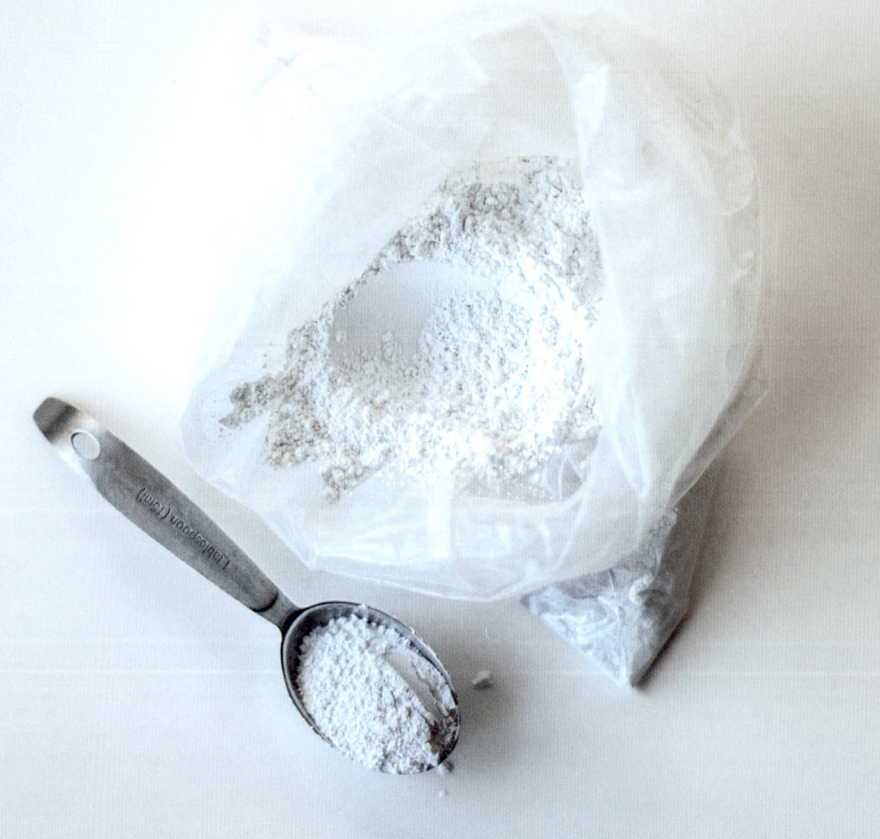

LIMEWATER

When making lake pigments, calcium hydroxide is used in solution. It's very sparingly water-soluble, so just a small amount is needed for a saturated solution with a fine dusting of white precipitate.

MATERIALS

· 2g calcium hydroxide

· 1L distilled water

METHOD

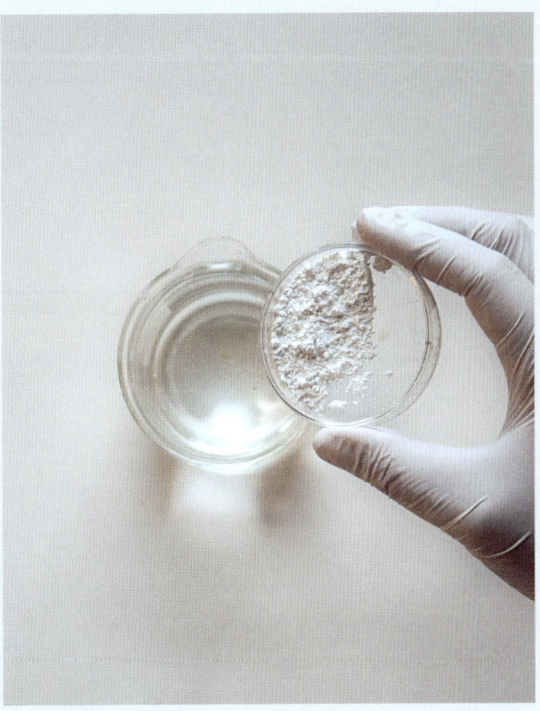

1. Fill a jar or beaker with 1 liter of distilled water. Sprinkle calcium hydroxide over the liquid, and then stir. Rest overnight.

2. To use limewater, decant the clear liquid without disturbing the sediment.

When laking with limewater, use pH indicator strips to pinpoint the correct amount.

Note: Limewater will leave a misty white rime of crystals on your tools. Don't bother scrubbing; it dissolves easily in vinegar.

Wood Ash Water

Wood ash water is just what it sounds like: the liquid produced by soaking hardwood fire ashes in water. The water infuses with soluble alkali salts from the ashes, primarily potassium carbonate. If you were to simmer down great quantities of wood ash water, you'd be left with a pot full of potash crystals— the origin of the word "potassium." Further heating potash to drive off impurities refines it to pearlash. These ingredients appear in historical dye recipes and craft manuals, functioning as important and relatively accessible alkalis. Dyers, artisans, and alchemists called wood ash water "lye," but today we reserve that term for the more potent sodium hydroxide. Modern lye is overkill for pigment-making purposes, so to head off any confusion, I'll avoid this term and stick to wood ash water.

When laking with wood ash water, use pH indicator strips to pinpoint the correct amount.

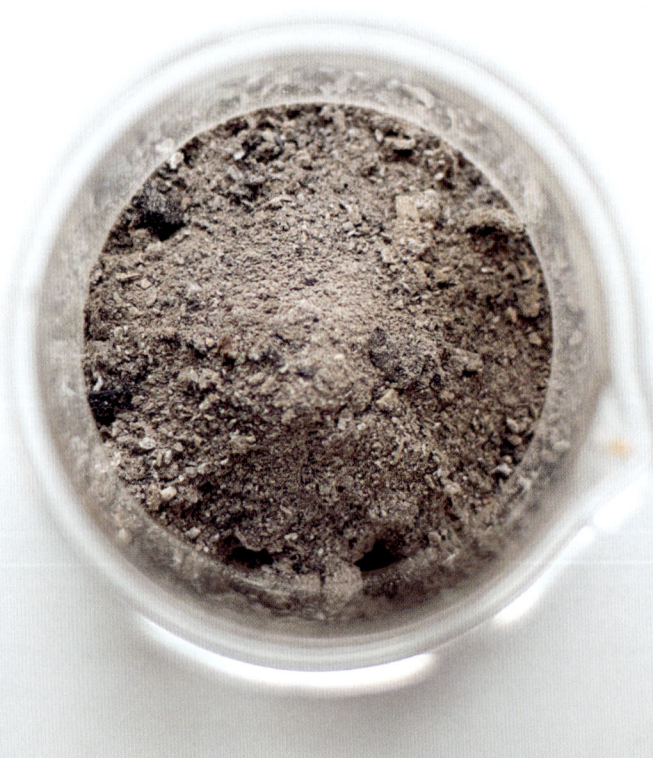

WOOD ASH WATER RECIPE

For this mixture, use fresh (but cold) ashes from oak, beech, ash, or other hardwood trees.

MATERIALS

· Roughly 2 cups hardwood ashes

· Distilled water, rainwater, or soft water

METHOD

1. Fill a wide mouthed glass container three-quarters full of water.

2. Wearing a face mask, scoop ashes into the water, filling up to half the volume of the container with ashes. Top up with additional water if needed.

3. Allow the ashes to settle, and steep for one to two weeks. If any black cinders float to the surface, wear gloves to scoop them out with a small sieve.

4. The wood ash water is ready when it has a pH of 11 or higher and feels slimy when rubbed between gloved fingers. To use it, decant the clear liquid without disturbing the sediment; it might be helpful to use a ladle.

5. When the liquid has been used, compost the ashes and start a fresh batch.

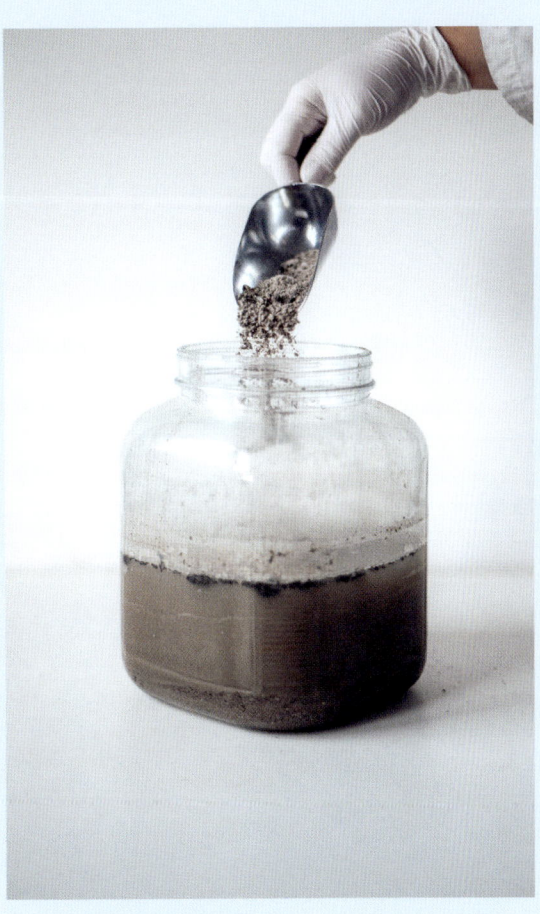

Chalk

Calcium carbonate occurs widely in the natural world; limestone, marble dust, seashells, and even cuttlefish bones are composed of it.

Chalk white is a natural pigment made of calcium carbonate, which dries opaque in water-based media but is translucent in oil and wax binders. Chalk can also be found in pastel sticks and contributes to their characteristic texture.

Chalk is an insoluble base and acts as a buffer to the presence of acid. This means it can replace soda ash in any of the lake pigment recipes (with the exception of the Alkaline Extraction [page 110], which requires an alkalizing agent rather than a buffer). Chalk donates carbonate anions until the alum solution neutralizes and lake pigment precipitates; it then contributes leftover insoluble gypsum (calcium sulfate) and any unreacted calcium carbonate. This combination of white powders (both earth pigments in their own right) provides body and opacity, stretching the pigment and making it easier to grind when dry. Chalk-neutralized lakes are thus a heterogeneous blend of colored aluminum hydroxide, gypsum, and chalk.

In laking, chalk is used at 50 percent the weight of alum (WOA) or enough to neutralize the dye/alum mixture. More chalk will not damage the pigment; it will simply result in a paler shade. Because chalk is so sparsely soluble, there is no need to dissolve it before adding it to a dye/alum mixture. Simply sprinkle chalk over the dye, and gently stir. It is normal to observe a pale stratum of gypsum precipitate before the aluminum-based lake. The yield of chalk-neutralized lakes is roughly 105 percent WOA or higher, considerably greater than those made with soda ash.

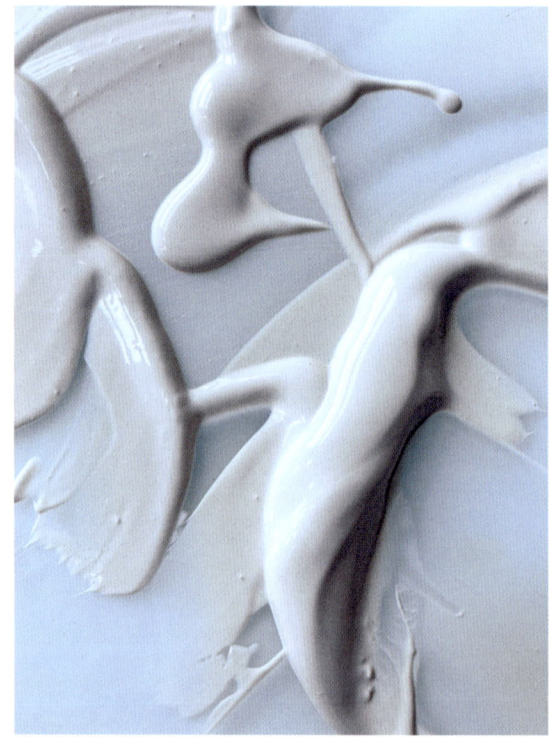

When compounding a watercolor, a sprinkle of chalk transforms the paint into a rich, creamy opaque gouache.

EGGSHELL WHITE PIGMENT

This chalk alternative is a wonderful white pigment for paint or pastel (warm, cool, or neutral depending on the color of your eggs). Eggshell powder can also be used in laking as a source of calcium carbonate instead of chalk.

MATERIALS

· 4 to 8 eggshells, any color

· White vinegar

· Distilled water

· Mortar & pestle

· Muller & slab (page 164)

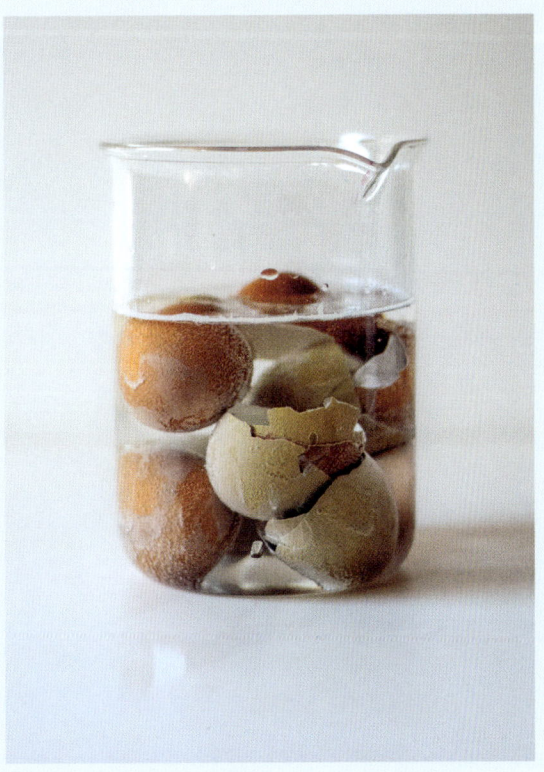

METHOD

1. After enjoying your eggs, gently rinse their shells with tap water. Keep the eggshell halves as intact as possible, and stash them in a safe place until you have four to eight in total.

2. Cover the eggshells with equal parts vinegar and tap water. Soak for 10 to 30 minutes, but do not leave unattended indefinitely, or the shells will gradually dissolve.

Recipe continued on next page.

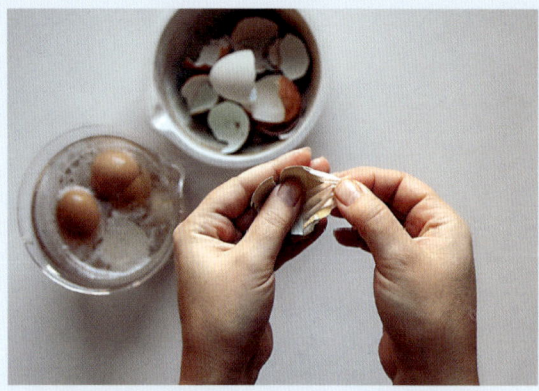

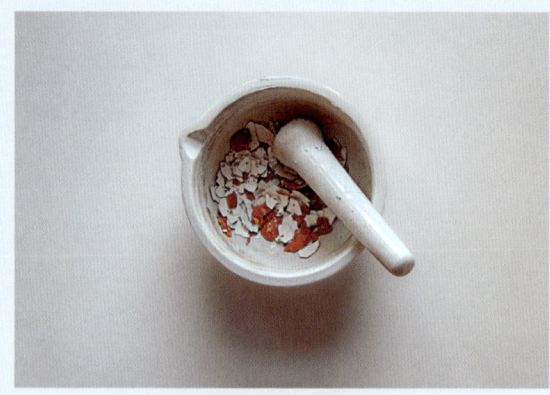

3. When the membranes pull easily from the inside of the shells, gently remove each one and discard. If your eggs are fresh from the henhouse, you may be able to skip the vinegar soak and peel the membranes out straightaway.

4. Place the eggshells in a mortar and pestle, and spritz them with water to keep the dust down. Calcium carbonate does not gather into hard chunks as it dries, so I like to grind and mull eggshell white as a wet paste. You can step away at any point in the grinding process; when you return, the dry eggshell powder can be easily broken apart and rewetted to continue grinding.

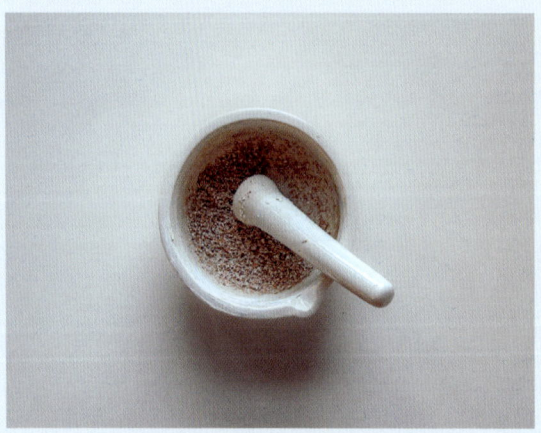

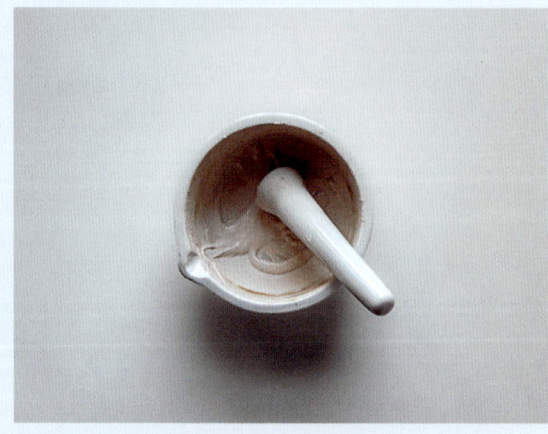

5. Crush and grind the shells as finely as possible in the mortar and pestle. This will require a bit of patience; eggshells are much sturdier than they seem. Alternatively, grind the shells in an artmaking-dedicated coffee mill or spice grinder.

6. Continue this process until the eggshell mixture reaches the consistency of toothpaste.

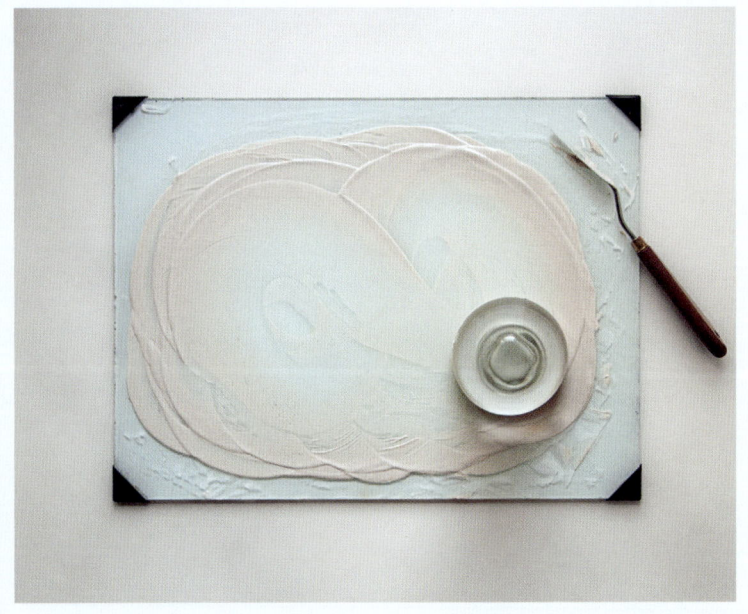

7. Transfer the eggshell paste to a slab. Continue processing with a muller, adding water as needed to keep the muller suctioned to the slab, until smooth. Again, this requires patience as the particles gradually diminish (see pages 50–51). You may step away for hours or days, and then rewet to continue mulling.

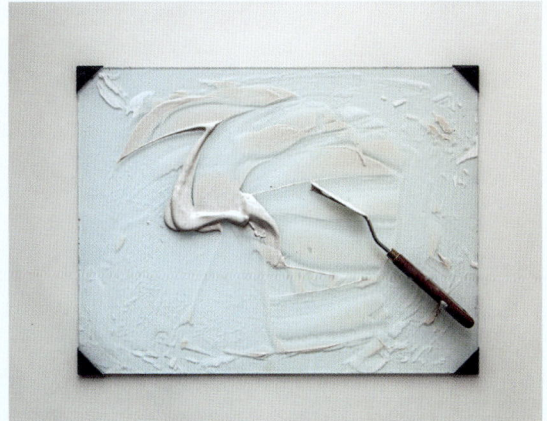

8. Using a palette knife, periodically scrape pigment from the sides of the muller and collect it from the edges of the slab. Rub a bit of the pigment between your finger and thumb. The eggshell white is ready when it is impalpably fine and appears similar to buttercream.

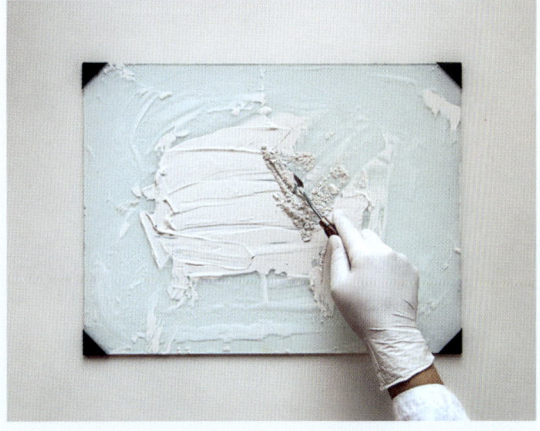

9. Allow the pigment to dry fully, scrape into a sealed container, and store indefinitely.

As the eggshell white is mulled, its visible, audible,
and palpable texture softens.

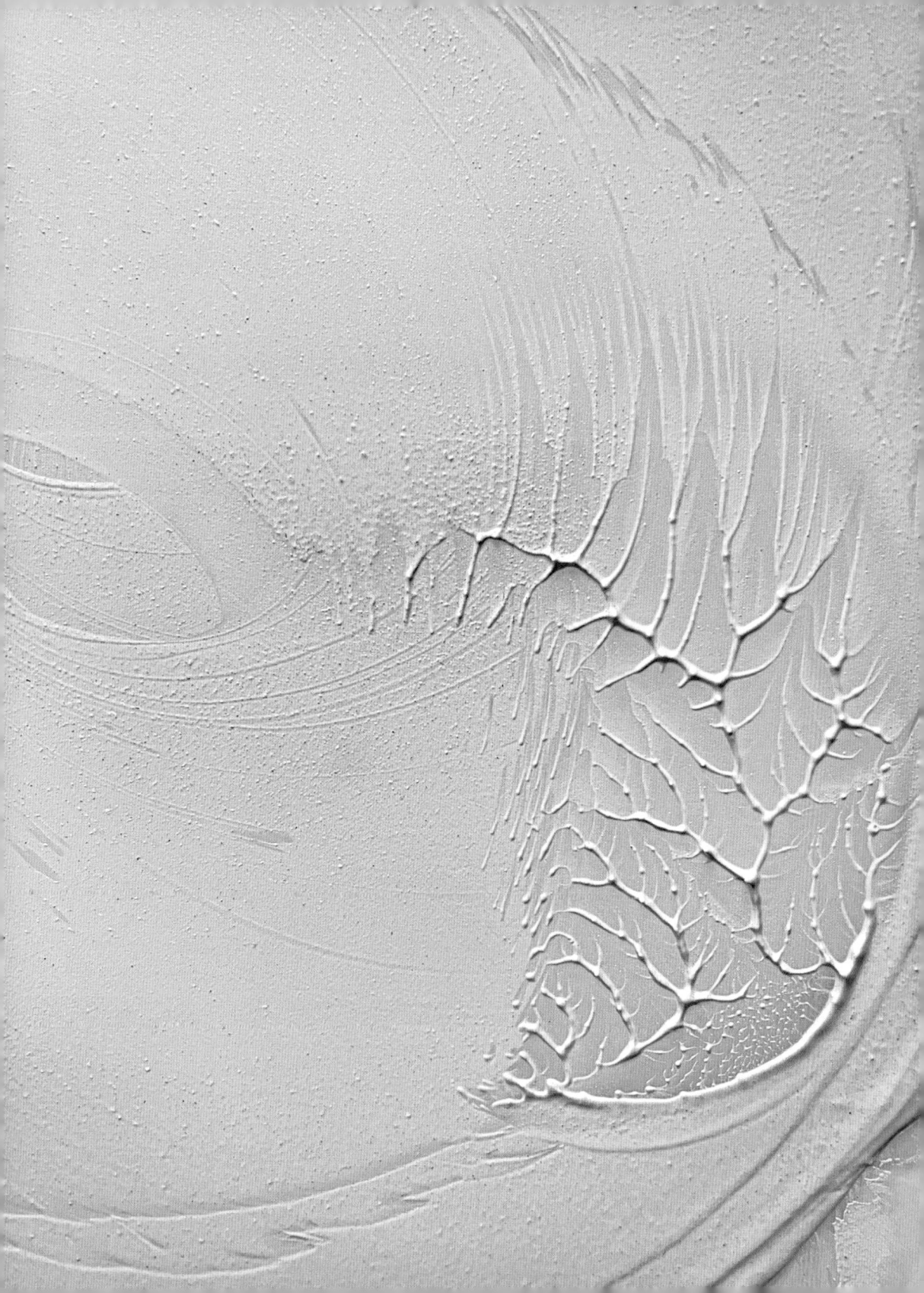

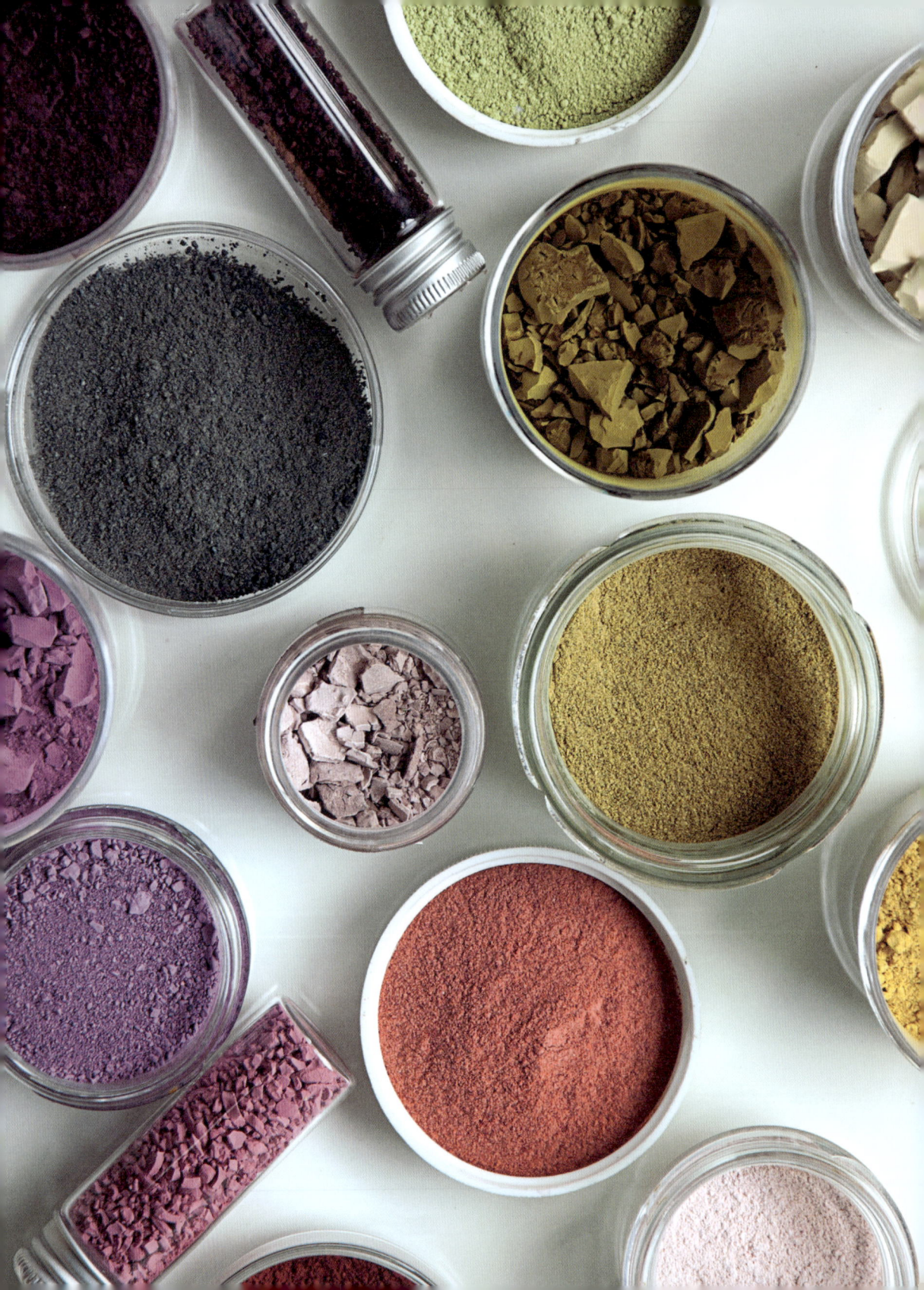

3

LAKE
PIGMENTS

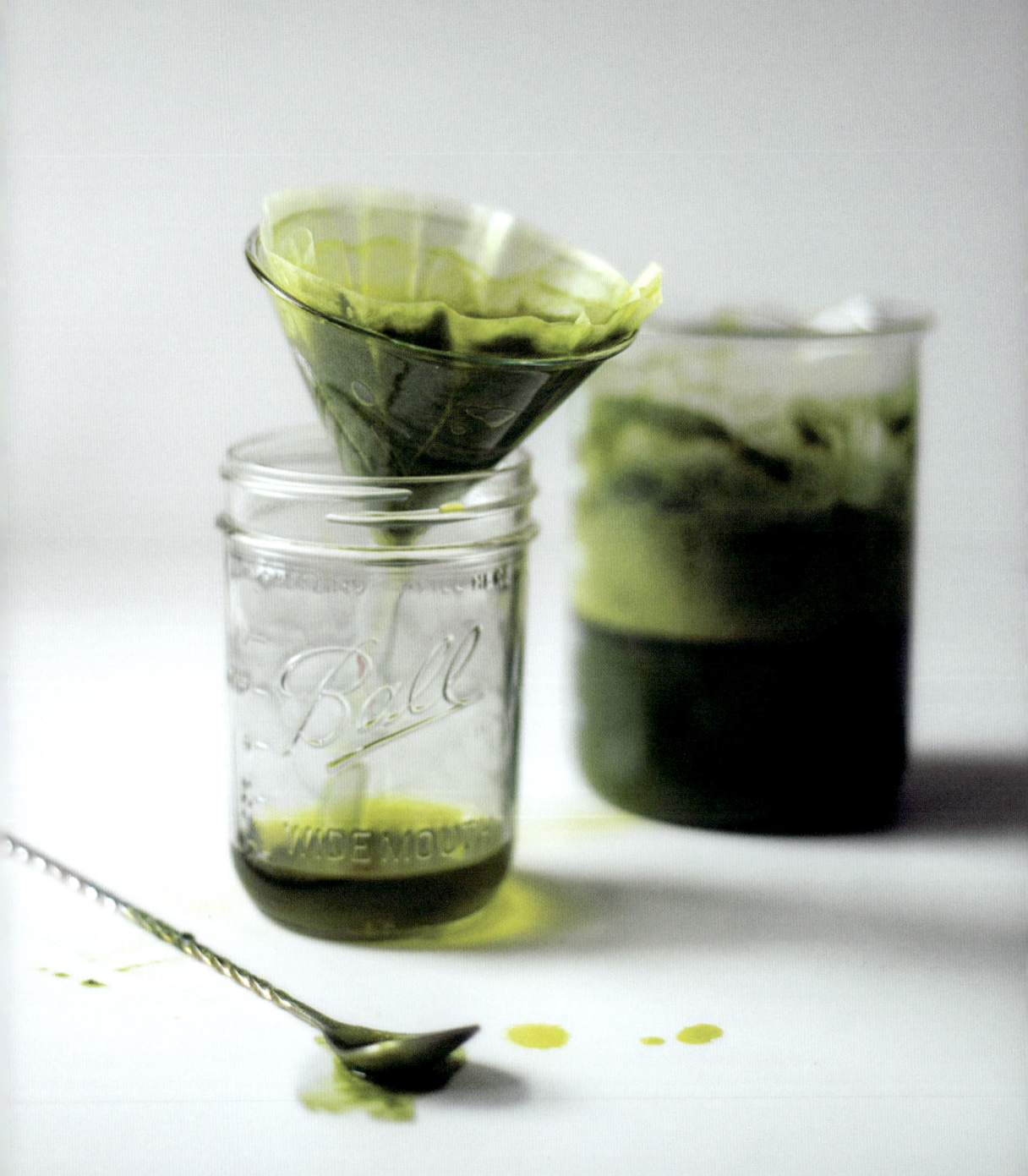

Lake pigments are simply botanical dyes affixed to an aluminum carrier. They can be made in a broad spectrum of hues, all following the same process. This method is shared in the Master Lake Pigment Recipe in this chapter, followed by a Dyestuff Ratios Chart for applying the process to a variety of plant sources. Together, these tools allow you to explore the pigment possibilities of both common dye plants, and new botanical acquaintances. A bit of light chemistry, along with hands-on practice, empowers you to tackle alternative extraction methods and color modification, extending the depth and breadth of the botanical palette.

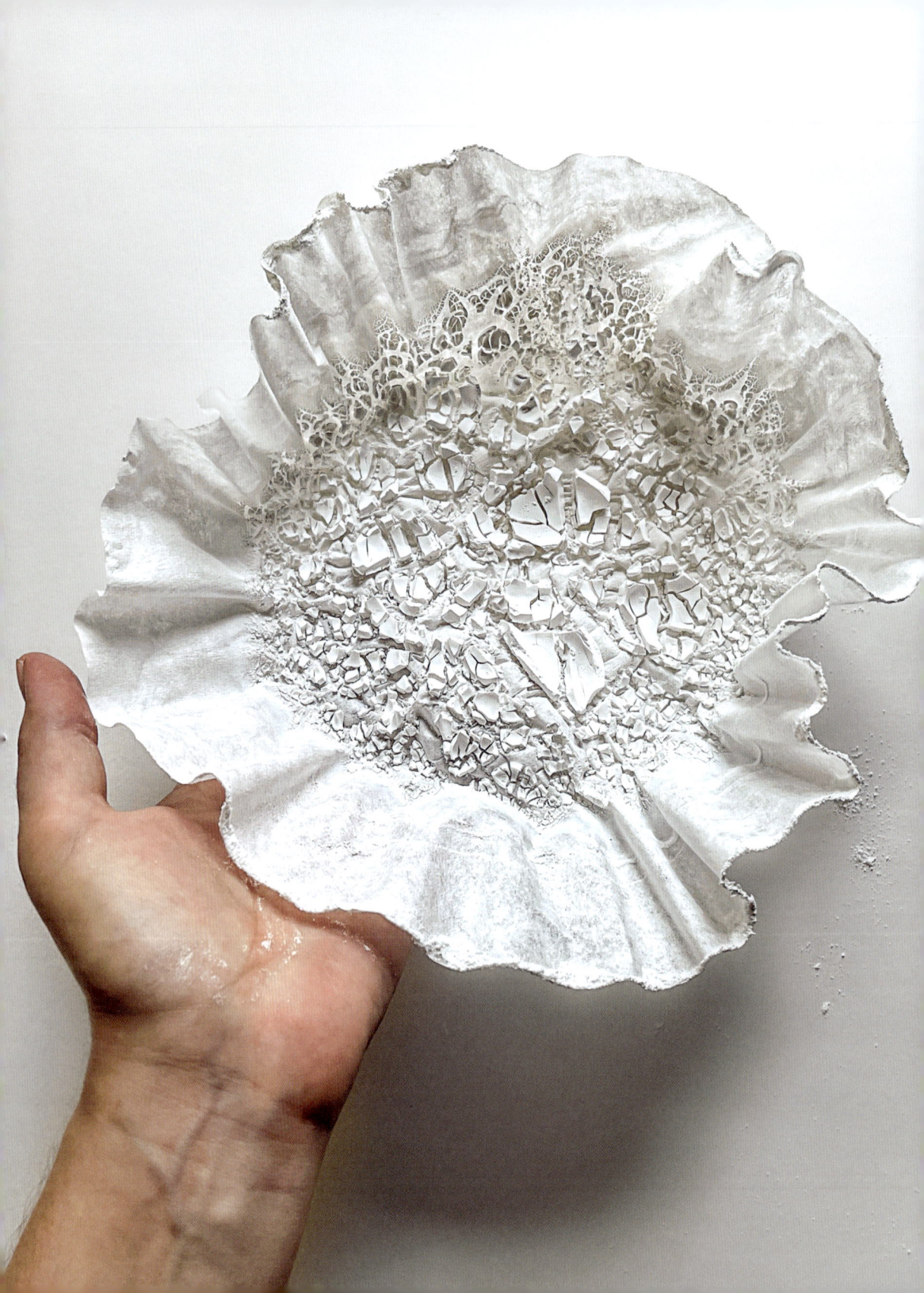

Aluminum Hydroxide

The splendid and varied colors of botanical dyes are water-soluble. These filmy, translucent hues are given substance when affixed to a mineral carrier, creating a lake pigment. That pigment is no longer soluble in water, extending the use of botanical dyes from aqueous applications into the realm of paints, pastels, and artists' materials of many sorts.

Aluminum hydroxide, also known as hydrated alumina, is the mineral core of, and bulk of, a lake. On its own, it's a transparent white pigment with little visual appeal; however, it's an ideal carrier for botanical colors because it fortifies them against damage from sunlight exposure. Having no color of its own, it brightens and stabilizes botanical hues.

Aluminum hydroxide is made from an ingredient familiar to natural dyers, which has an intrinsic chemical attraction to dyes: the mordant alum (page 37). As you'll recall, mordants are mineral salts with the friendly chemical propensity for attaching to other molecules, such as dyes and textile fibers. In fact, natural dyers create lake pigments each time they color fabric with a mordant dye, precipitating color into union with both mordant and fiber. The recipes in this section follow a similar process (simply leaving fibers out of the relationship).

To transform water-soluble alum into insoluble aluminum hydroxide, an alkali is needed to offset the inherent acidity of the alum. The neutralization reaction between the acid and base causes an effervescence of carbon dioxide. Beneath the bubbles, the colorful, translucent liquid dye fills with a cloudy mass of particles when the aluminum hydroxide materializes. It then precipitates out of solution, carrying the dye color with it. The lake pigment simply needs to be filtered, rinsed clean, and dried.

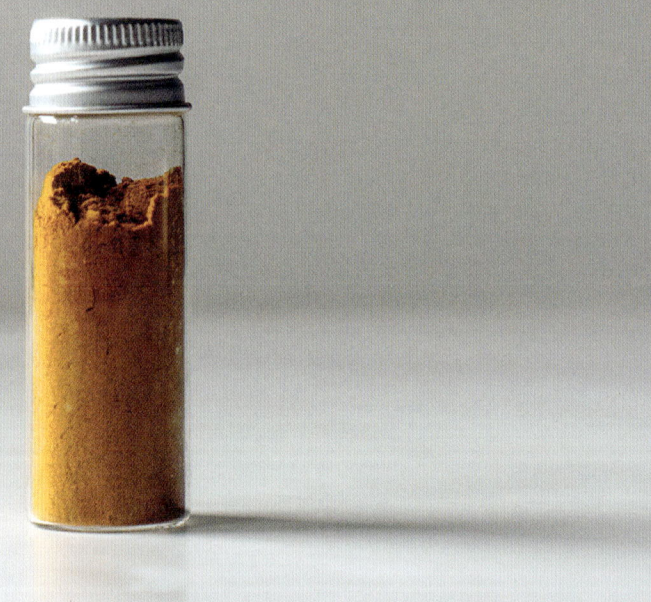

MARIGOLD BLOSSOM LAKE PIGMENT

Master Lake Pigment

The pigment-making neutralization reaction requires just four ingredients: dye, alum, base, and water. Let's have a quick review of these four ingredients.

Dye

Any dye that bonds to mordants can be used to create lake pigments. The great majority of natural dyes are called "adjective" or "mordant dyes" (page 37) and include colors from plants, insects, and some mushrooms. If you venture away from mordant dyes and into botanical possibilities, you might find that some plants leave a lot of color dissolved in solution and little on the aluminum hydroxide pigment. Some dyes are weakly attracted to alum, and because multiple colorants exist comingled in each plant, results when laking may be surprising. The only way to determine what color a new plant will produce is to make an experimental sample. A separate class of natural dyes are the vat dyes, including indigo and woad (page 133). They hold no chemical affinity for mordants. These beautiful blues function as pigments without laking.

Alum

Alum will contribute the core and bulk of the pigment. Because of this, all the calculations in laking are percentages relative to the weight of alum, or WOA. Each of the following recipes calls for 10 grams of alum, with a proportional measure of dye material, water, and base. With this system, the recipes can be easily scaled up or down.

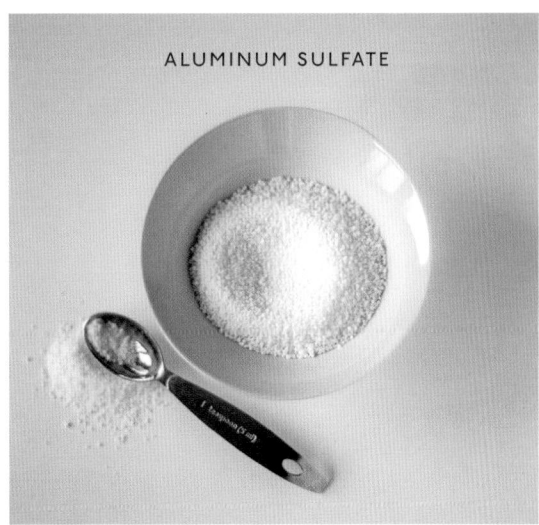

ALUMINUM SULFATE

Base

Many different ingredients can offset the alum's acidity. Soda ash (sodium carbonate) is a readily available base that produces translucent pigments. The amount of base called for is just enough to neutralize the mixture and can be determined by the old-fashioned method of simply listening for fizzing carbon dioxide bubbles while slowly adding the soda ash. When the fizzing ceases, the neutralization is complete and enough base has been added. Conveniently enough, soda ash predictably precipitates a lake at 50 percent the weight of the alum, no listening required.

Base substitutions (pages 41–51) allow you to vary your pigments' hiding strength, making them opaque or semiopaque. Some also make use of home and kitchen waste, turning fireplace ashes and eggshells into thrifty, hyper-local ingredients.

Water

Water is the environment in which our ingredients mingle, react, and create a lake pigment. We need only enough water, at a warm enough temperature, to ensure that everything soluble is dissolved. At the laking stage, all ingredients should all be heated to between 43 to 76°C (110 to 170°F). Hot tap water, with a stir, is sufficiently warm to dissolve alum and soda ash.

A recipe calling for 10g alum should be laked in roughly 500ml water. A smaller volume of water makes for a concentrated, vigorous neutralization reaction, potentially leaving unreacted ingredients in the froth. A larger volume of water will still allow a lake pigment to form, but the diffuse neutralization reaction might not produce discernible fizzing. Even in a vast stewpot of water, the ionically charged molecules will find each other, and a lake pigment will ultimately be captured by your filter paper.

Once a lake pigment is formed and filtered, you must wash it to remove mineral salts and residual dye, preventing crystallization and spoilage. I recommend distilled water for washing lake pigments.

TIME & MEASUREMENTS

The active time required to complete the Master Lake Pigment Recipe is about one to two hours, but the entire process stretches over a few days. Much like homemade bread dough, a lake pigment must be minded and moved along at key moments, and the steps will become instinctual with practice. The time commitment ultimately rests on the quantity of pigment to filter and the weather conditions that affect drying time. Choose a time when you can periodically check on your pigment's progress, topping up filters, and moving from one step to the next. For ease of mathematics, I've used the metric system throughout the recipes. It's worth noting that one milliliter of water weighs one gram, meaning we can easily measure 100ml as 100g on a digital scale. For the success of your pigments, take great care if you choose to convert the small metric measures to imperial.

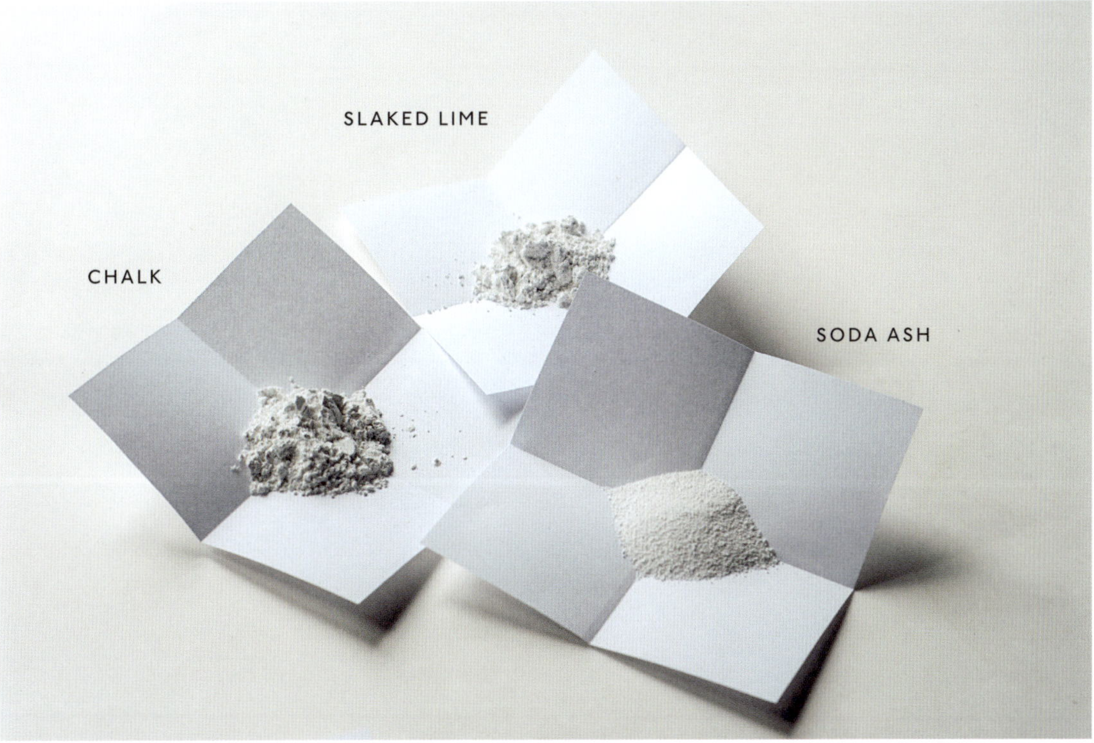

CHALK SLAKED LIME SODA ASH

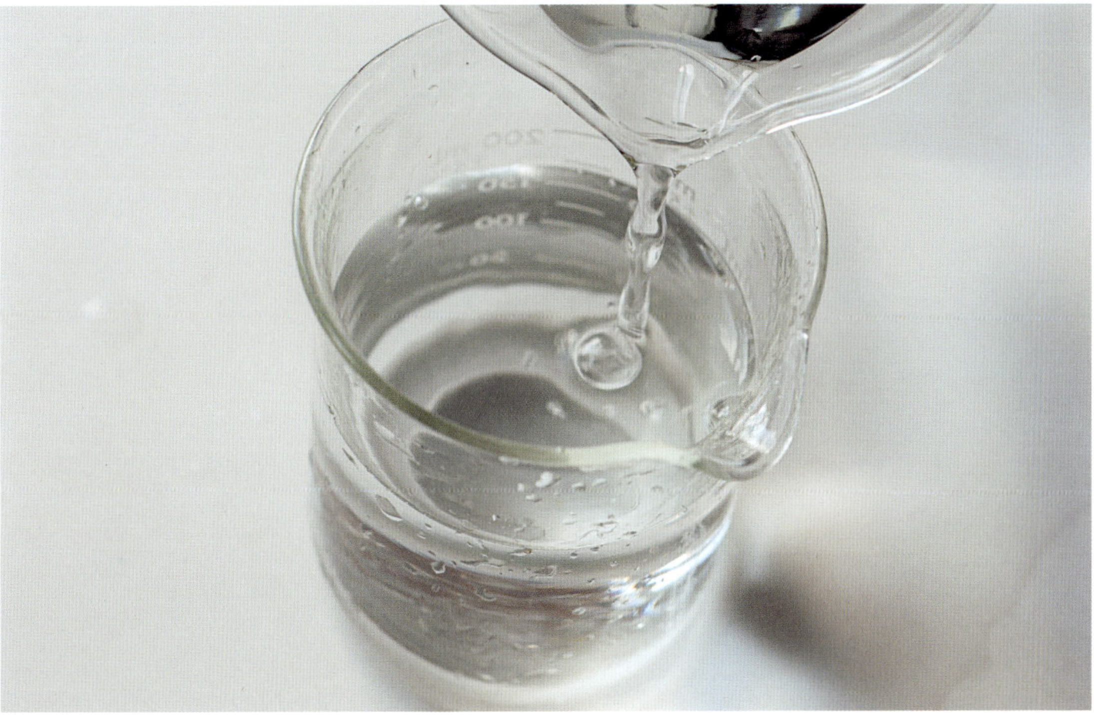

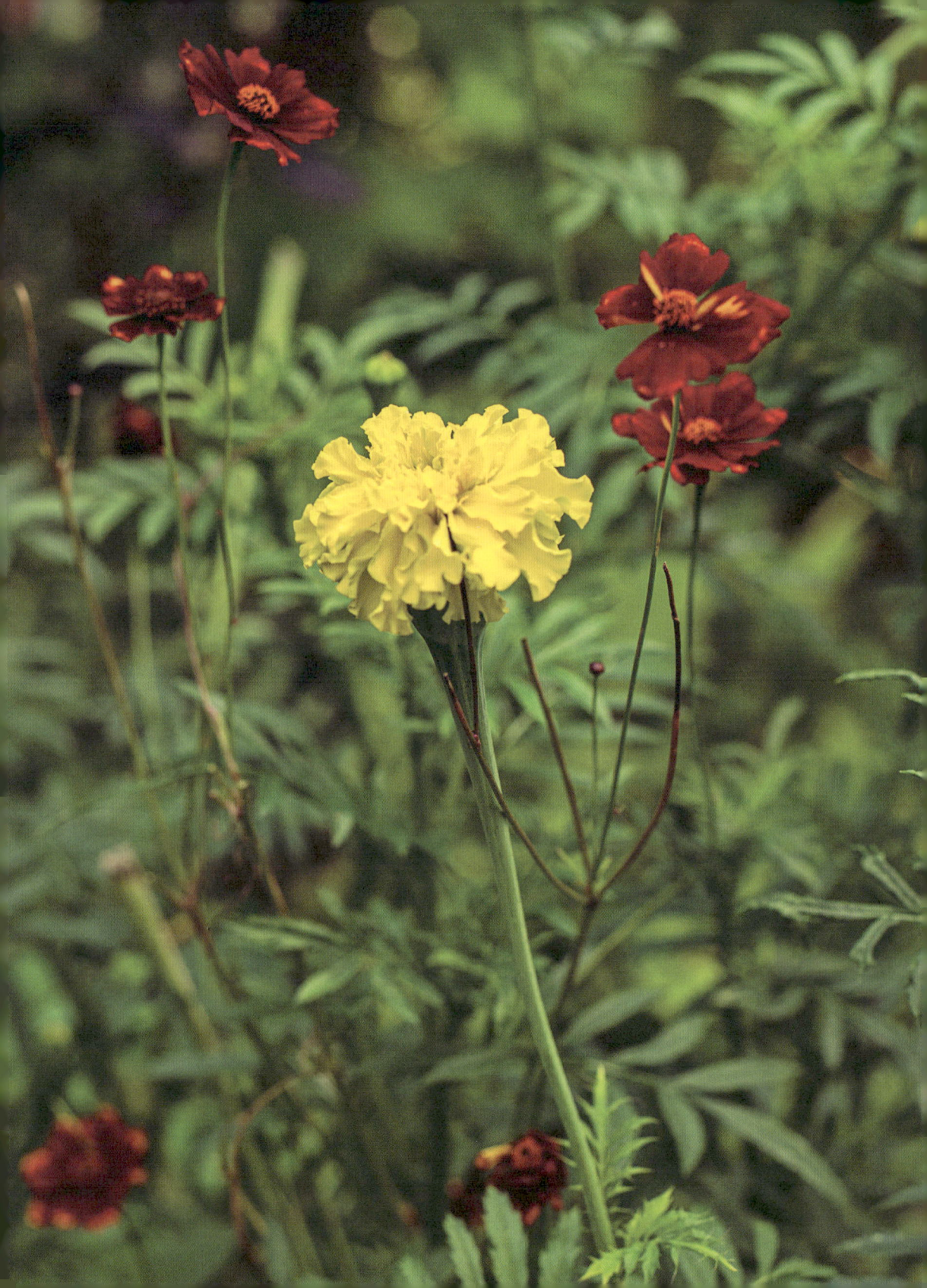

MASTER LAKE PIGMENT RECIPE

This is my everyday laking process, and it's the recipe I use to test unfamiliar dyes and potential color-bearing plants. I've used marigold blossoms for recipe demonstration purposes.

MATERIALS

· Dye material, measured in proportion to alum (see page 70)

· 10g alum (aluminum sulfate or potassium aluminum sulfate)

· 5g soda ash (sodium carbonate)

· Distilled water for washing pigment

METHOD

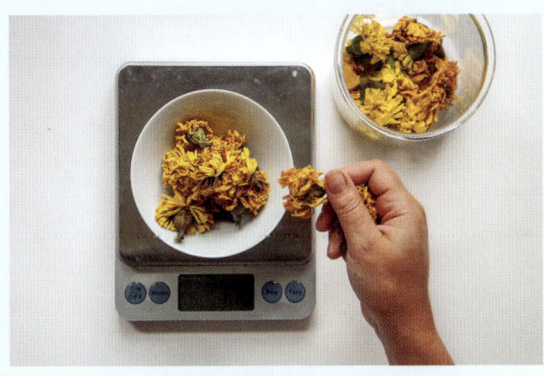

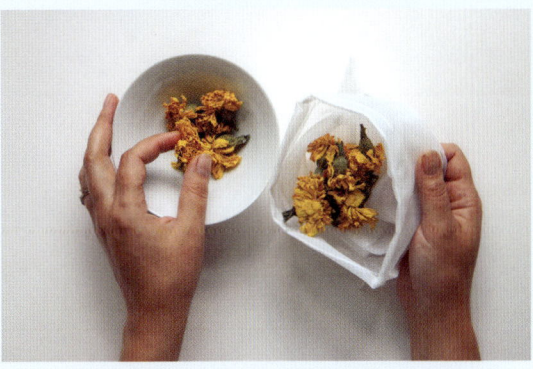

1. Using the Dyestuff Ratios chart on page 72, calculate the weight, and measure out dry plant material relative to 10g alum. Because the chart calls for 200% marigold, I am starting with 20g dried marigold blossoms.

2. Optional: If your dyestuff is finely ground, enclose it in a fabric or mesh sachet for extraction. This will contain most of the particulate but will still allow the dye to seep out.

Recipe continued on next page

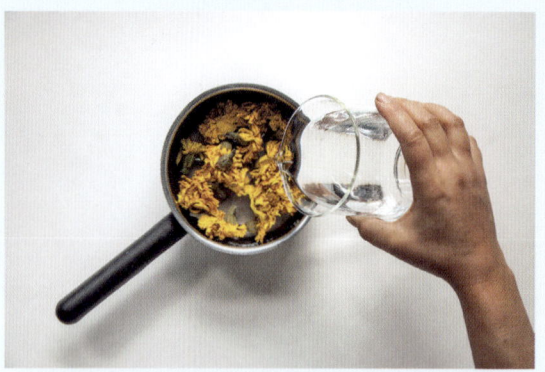

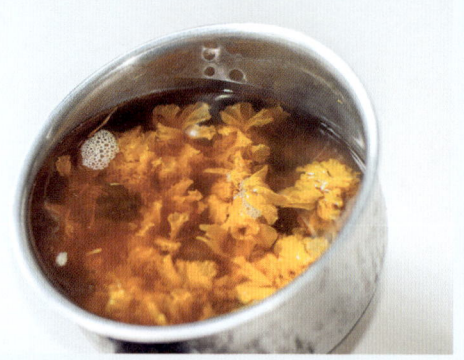

3. Cover dyestuff with about 750ml water and bring to a simmer for 30 minutes to two hours. Delicate petals and leaves need only a short time to release their dye, while hearty roots and wood benefit from a longer one- to two-hour extraction. A thorough decoction (extracting dye via simmering water) will provide the greatest amount of color.

4. Continue to top up the water level as needed to keep the plant material submerged during extraction.

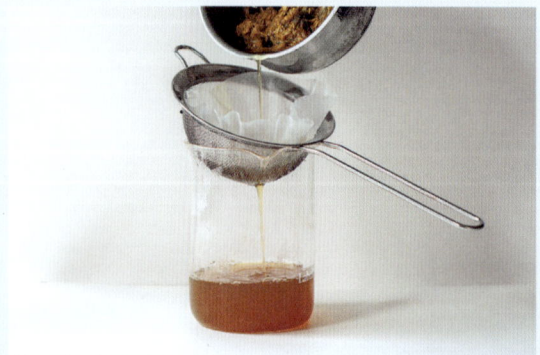

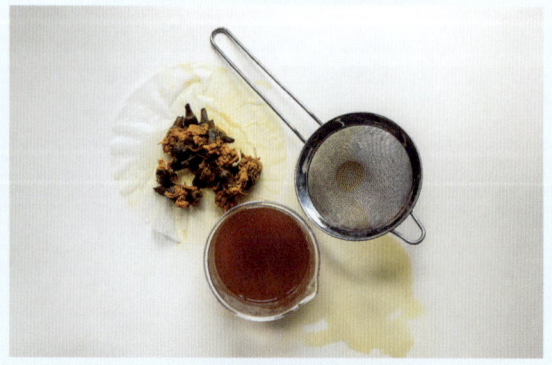

5. Pass the liquid through a coffee filter to remove all plant material.

6. Compost or discard the spent dyestuff.

Recipe continued on next page

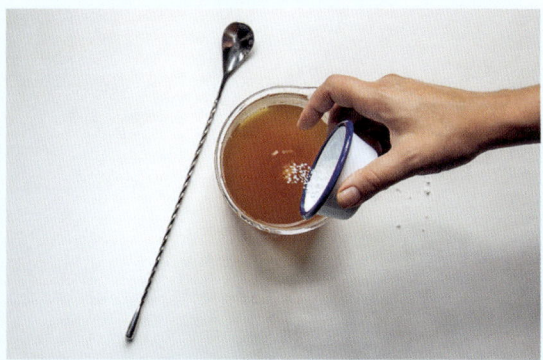 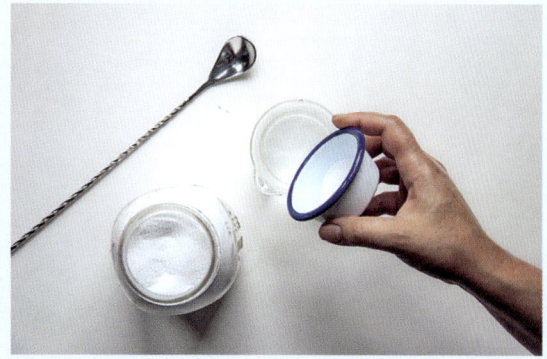

7. Transfer the liquid to a 1-liter container or larger. It is helpful to use a clear container, such as a wide-mouthed canning jar, to observe the chemical reaction unfolding. Because the neutralization reaction will produce effervescence, the container must be two to three times larger than the volume of dye. The quantity of carbon dioxide bubbles can be surprising—and even create volcanic overflow! Add hot water to the dye to equal 500ml if needed. The temperature should be between 43 to 76 °C (110 to 170 °F). Add 10g alum to the hot dye; stir to dissolve.

8. In a separate container, stir 5g soda ash into about 200ml hot water.

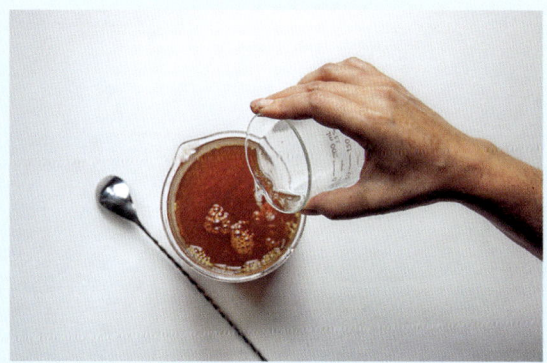 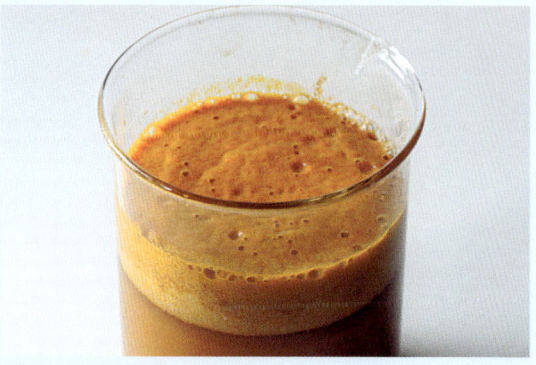

9. Slowly pour the soda ash solution into the dye mixture, stirring gently.

10. Expect frothing when the acid and alkali meet! If it looks like it's going to overflow, use a spray bottle to spritz the froth with water. The translucent liquid will turn opaque as pigment particles coalesce.

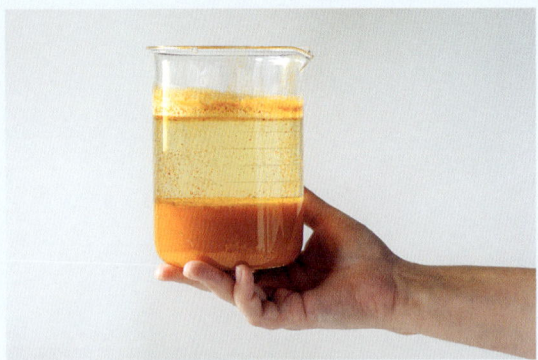

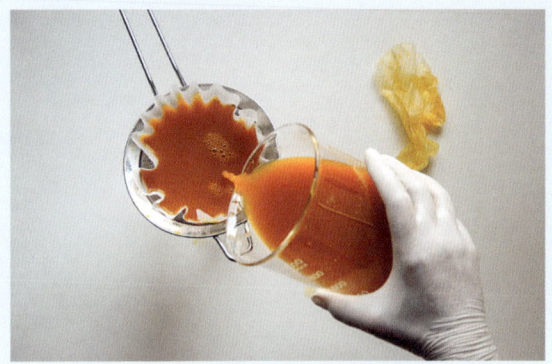

11. Optional: Allow the pigment to settle for one to 12 hours. This allows you to see the colorful solids gradually drift to the bottom of the vessel, with the translucent liquid above. The leftover liquid is known as the supernatant. It contains residual solutes, including salts, any possible superabundance of dye, and any plant colorants not attracted to alum.

12. Wait five minutes after adding the soda ash for the pH to stabilize and the pigment to form. Place a coffee filter in a sieve or funnel over a clean jar or beaker. Pour the lake pigment mixture into the filter, topping it up in stages if necessary. It might take several hours for all the liquid to wend its way through the superfine particles, but gravity will win out. Don't dispose of any foamy froth—heap it into the filter too; it will be reincorporated during washing.

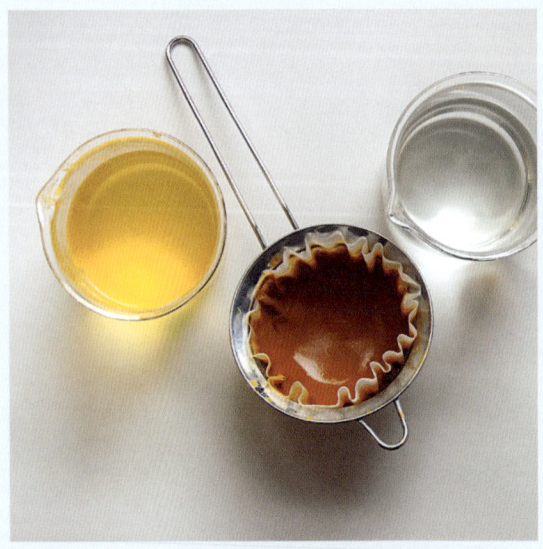

13. The colorful pudding-like paste retained by the filter is the lake pigment. Optional: If the supernatant liquid that passes through the filter is still richly colored, you may choose to lake it to make another, paler pigment. Reheat, and add 5g dissolved alum followed by 2.5g dissolved soda ash.

14. The lake pigment must be washed to remove any lingering dye that might cause fading, staining, or spoilage, as well as residual salts, which will crystallize. Pour roughly 500ml distilled water into a clean jar.

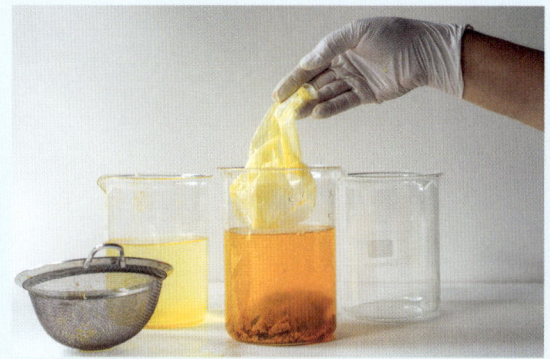

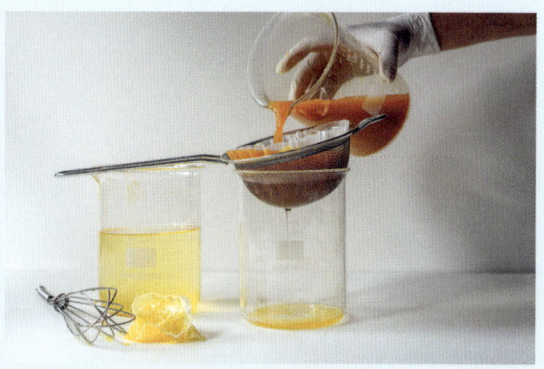

15. Wearing gloves, transfer the entire filter paper full of pigment into the distilled water. Swish the paper around to release the pigment into the water, lift the paper out, and discard.

16. Gently whisk or stir the mixture to break up any clumps of pigment, and let it rest for five minutes. Filter. During the first 30 seconds, a small amount of pigment might seep through the filter with the liquid. Return this liquid to the filter to avoid losing any pigment and to accurately judge the color of the rinse water. Repeat washing and filtering once more, or until the pigment sheds no dye and the rinse water is close to clear.

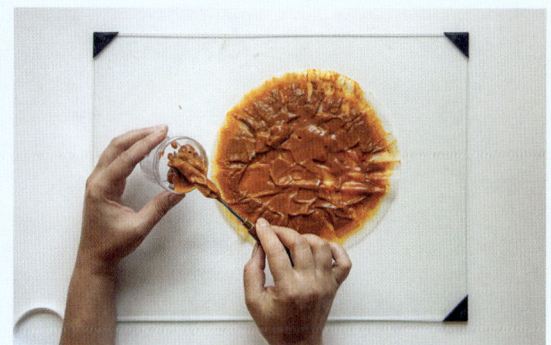

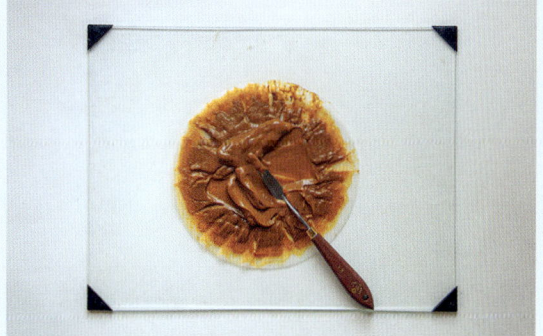

17. Optional: Store the wet pigment paste indefinitely in a lidded jar in the refrigerator, to be used for Speedy Paint & Ink (page 151).

18. Spread the filter paper of pigment out to dry, ideally in a warm spot out of direct sunlight. If you wish to hasten drying, do not heat higher than 60°C (140°F). I use a heated seed germination mat or studio-dedicated food dehydrator if needed.

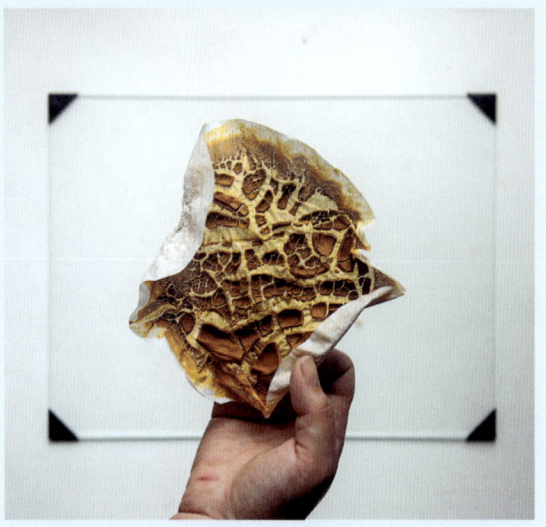

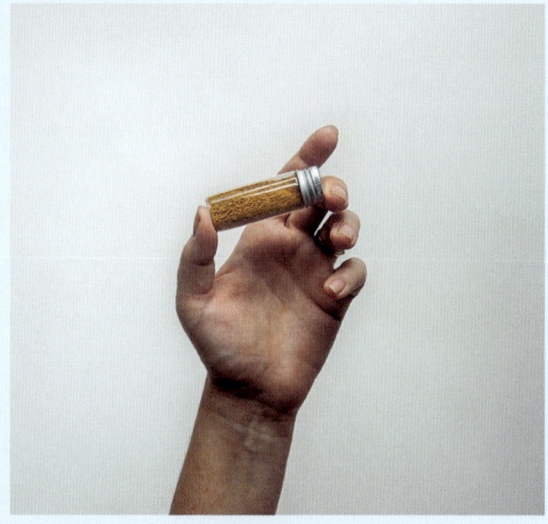

19. As the pigment dries, it will gather into clumps and its shade may shift.

20. Store dry pigment indefinitely in a lidded container out of direct sunlight.

YIELD & SCALING UP

The aluminum hydroxide precipitated by laking is equal to about 35 percent of the weight of alum (WOA), plus the weight of dye. A lake made with 10g alum provides 3g to 6g pigment, or 1 to 2 tablespoons. This is enough for a few pans of watercolor or pastel sticks. The whole undertaking can be accomplished with modestly sized tools: a small saucepan, 1-liter canning jars, a funnel or small sieve, and standard 20cm (8in) coffee filters. If needed, the recipe can be scaled down to as little as 2g alum with commensurate yield.

Because laking is a lengthy process (and I like to have enough pigment to really splash out), I often make larger batches. If you increase the recipe to more than 10g alum, scale up to a 4-liter dye pot, 5-liter plastic masonry mixing buckets, 20cm (8in) stainless steel sieve, and 33cm (13in) barista-worthy coffee filters. If your art demands more pigment, you can lake in a 20-liter (5-gallon) bucket with a 25-micron paint strainer insert serving as filter.

Troubleshooting

My filter is clogged, and the dye or supernatant isn't passing through. What can I do?

Allow 12 to 24 hours for as much liquid to pass through as possible. If that doesn't work, don't try to handle or squeeze the filter; it's likely to tear. Instead, delicately pour the liquid pooled above the solids into a fresh filter. Review step 5 for Weld Lake on page 93 for a workaround.

Can I filter through fabric instead of coffee filters?

Yes. You can use and reuse a very tightly woven fabric to filter your pigment. Wash the fabric between uses to prevent backstaining.

It seems wasteful to discard this plant material after extraction. Can I use it again?

It's not worth making a secondary extraction from plants that easily yield their color the first time; however, stubborn roots and wood are worth saving. You can make a second extraction immediately, or cache dry or frozen dyestuff for the future. When the time comes for laking, expect a paler color, or decrease the recommended amount of alum and alkali by half.

My rinse water isn't getting clearer.

Measure the pH of the lake mixture to make certain it falls between 6 and 8, and then adjust with a small amount of additional alum or soda ash, if necessary. Measure the pH of your water source, and switch to distilled water if it isn't neutral. If you are using a fabric filter, switch to a paper coffee filter, or try two nested paper filters.

Can I take a break in this process?

I find it most convenient to pause after laking but before washing. At this point, the dye is stabilized, so it's not prone to oxidization or spoilage. Cover the entire jar of lake pigment and supernatant, and refrigerate it for up to two weeks. Or, if you have filtered the lake, transfer the pigment paste to a lidded container and refrigerate it. Remember that it will still need to be washed when you to return to it. The growth of fine white crystals around the edge of the paste indicates the lingering salts that need rinsing away.

I've made a pigment, but it has almost no color. What happened?

This can occur when we experiment with unconventional dye sources. Some biochromes are simply not attracted to alum and just slide off the aluminum hydroxide, leaving it pale or completely colorless. Luckily, you can mount dye on fresh aluminum hydroxide paste. Simply make a new batch of dye from a mordant dye source, stir in the colorless pigment paste, filter, and wash the now-colorful lake pigment.

My supernatant has a lot of color in it! What can I do with it?

If you haven't already, try laking the supernatant to catch its color, using half the original amount of alum and alkali. If you still have lots of color in the solution, it's likely it isn't attracted to alum. You can try using this to dye scoured, unmordanted protein fibers.

Dyestuff Ratios Chart

The strength of dye sources is extremely variable; some plants are richly saturated with color, and some yield very little. This chart recommends a percentage of dye plant material in relation to the weight of alum (WOA) for making lake pigments. These percentages are a flexible guideline for strong color.

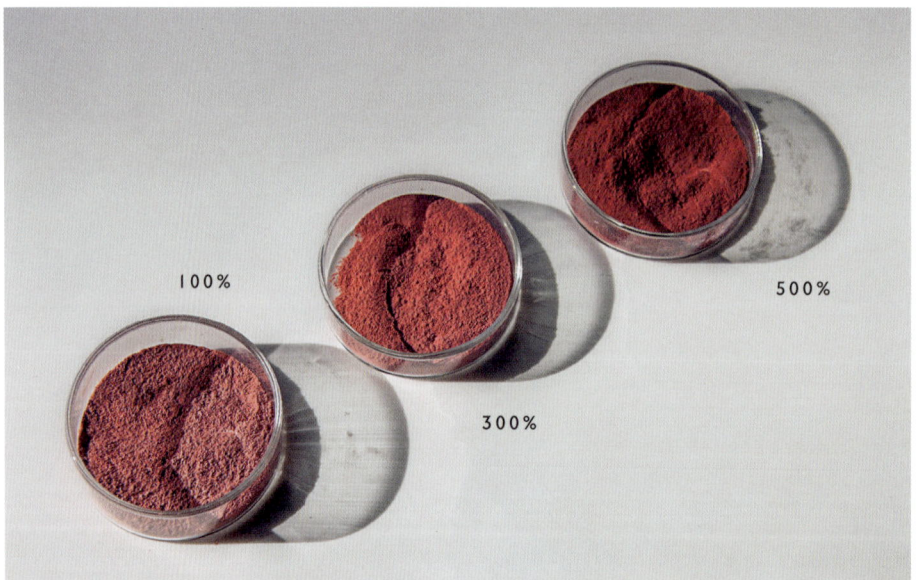

100%

300%

500%

Madder Root relative to WOA.

As the chart indicates, a strong dye source like logwood will call for minimal plant material—about one part logwood to two parts alum, or 50 percent logwood. A weaker dye source like madder requires much more plant material, say five parts madder to one part alum, or 500 percent.

FRESH VS. DRY

The plant sources listed are dried dyestuffs, unless noted otherwise. Fresh plants weigh more due to the amount of water they contain. Exactly how much more varies widely. To use fresh plants, multiply the percentages given by 3 to 6, judging the amount by how juicy the plant is. For example, I'd multiply the weight of fresh leaves by 3 and sappy berries by 6.

Using the Chart

Weight of Alum × Dye % = Weight of Dye

For example, to calculate the dry weight of goldenrod to lake with 10g alum:
10g alum × 1.8 = 18g goldenrod.

Recall from grade school that the decimal point moves two places left when converting a percentage to an integer, so 180 percent becomes 1.8.

If you'd like to start with a known weight of plant material, the equation is simply inverted:

Weight of Dye ÷ Dye % = Weight of Alum

Example:
18g goldenrod ÷ 1.8 = 10g alum

The percentages given are a guideline for richly valued pigments. If the amount of dye is decreased, its hue will be stretched across the colorless aluminum hydroxide, resulting in a paler pigment. This makes a perfectly functional pigment, but a pastel hue. With this understanding, we can intentionally create different tints from the same plant by varying the dye/alum ratio. The lower the percentage, the paler the pigment.

ALTERNATIVE EXTRACTION METHODS

Dyes are most commonly extracted in simmering water, leaching colorants into solution much like vegetables infuse stock with flavor. This is called "decoction" and is the method I follow in the Master Lake Pigment Recipe (see page 63). There are other ways to get dye out of a plant. Some colorants have greater solubility in alcohol than water, making a tincture the best option. Some colors develop well in an alkaline extraction, and some develop through fermentation. These alternative extraction methods and the dye sources best suited to them are listed below and detailed in the Lake Pigment Variations (see page 79). In addition to decoction, they bring further richness and nuance to the botanical palette. The alternative extraction methods also require less dyestuffs because they often elicit more color (an element of their allure).

EXTRAPOLATING TO OTHER PLANTS

Of the vast number of mordant dyes that can be laked, the ones listed on the chart are some of the most renowned and commonly available sources. If you would like to explore a plant not listed here, look for reputable guidance on the percentage used by dyers relative to the weight of fibers (WOF), and multiply that number by six. For example, if dried weld is recommended for dyeing at 20 to 30 percent WOF, make a lake using 120 to 180 percent weld relative to the WOA. The higher percentage provides a more intense value and, likely, a color-rich supernatant. If I cannot find dyers' guidance for a particular plant, I experiment with 200 percent WOA (that's 20g dried dyestuff for 10g alum) and go from there.

DYESTUFF RATIOS CHART

DYE SOURCE	DECOCTION (PAGE 63)
BIRCH BARK, *BETULA SPP.*	75%
BUCKTHORN BERRY, *RHAMNUS SPP.*	100%
BUCKTHORN LEAF, *RHAMNUS SPP.*	100%
BUCKTHORN BARK, *RHAMNUS SPP.*	75%
COCHINEAL, *DACTYLOPIUS COCCUS*	50%
DYER'S CHAMOMILE, *COTA TINCTORIA*	300%
DYER'S COREOPSIS, *COREOPSIS TINCTORIA*	200%
FRUIT TREE BARK (APPLE, PEAR, CHERRY, PLUM)	150%
GOLDENROD, *SOLIDAGO SPP.*	180%
IRIS FLOWERS, DARK PURPLE *IRIS SPP.*	100%
LAC EXTRACT, *KERRIA LACCA*	60%
LOGWOOD, *HAEMATOXYLUM CAMPECHIANUM*	90%
MADDER ROOT, *RUBIA CORDIFOLIA*	500%
MADDER ROOT, *RUBIA TINCTORUM*	500%
MARIGOLD, *TAGETES SPP.*	200%
MUGWORT, *ARTEMISIA VULGARIS*	150%
ONION SKIN, *ALLIUM CEPA*	120%
OSAGE ORANGE, *MACLURA POMIFERA*	180%
PINCUSHION FLOWER, DARK PURPLE *SCABIOSA SPP.*	100%
RHUBARB ROOT, *RHEUM SPP.*	180%
SANDALWOOD, *PTEROCARPUS SANTALINUS*	
SAPPANWOOD (EASTERN BRAZILWOOD), *CÆSALPINIA SAPPAN*	150%
SULFUR COSMOS, *COSMOS SULPHUREUS*	200%
TANSY, *TANACETUM VULGARE*	200%
TURMERIC, *CURCUMA LONGA*	250%
WELD, *RESEDA LUTEOLA*	120%

FERMENTATION (PAGE 120)	TINCTURE (PAGE 116)	ALKALINE EXTRACTION (PAGE 110)	ICE MACERATION (PAGE 104)
		40%	
		30%	
		80%	
	100%		600% FRESH
	50%	50%	
300%			
300%			
	100%		600% FRESH
		100%	
	500%		
	75%	75%	
	50%		

Lightfast & Fugitive

Plant colors serve a host of functions, but durability to light is not necessarily one of them. In nature, a biochrome, or natural colorant, simply doesn't need to last longer than the lifespan of its botanical host. Exposure to sunlight can cause chemical decomposition of the biochrome, thereby sapping its vibrancy.

Colorants that fade quickly are known as fugitive, and those that retain their color over time are lightfast. Alum bolsters lightfastness by stabilizing dye molecules—it's our best tool for staving off decay. Fortunately, alum is at the heart of every lake pigment, but even so, botanical pigments are not completely permanent. It's no surprise that lake pigments' historical uses included manuscript illumination and paper marbling, meaning the pages bearing those colors were safely tucked inside books where they'd be safe from sunlight.

Understanding the chemistry of handmade materials will help you know when to bend the rules or depart from a recipe for artistic purposes. The reactivity of botanical colorants invites creative intercession. Anthotypes are alternative photographs using dye as emulsion, reframing "fugitive" as "photosensitive." Not all dyes lighten with sun exposure; tannins noticeably darken. Consider the generative ways you might utilize changeable colors.

BEST PRACTICES FOR LONGEVITY

- Store dyestuffs and pigments out of sunlight.
- Protect artwork from direct sun exposure.
- Apply multiple coats to build up a density of color that is more resistant to light and fades less.
- Note that iron modification improves longevity.
- If lightfastness is a concern, use only the most durable dye sources: weld, madder, lac, and indigo. These can be mixed together for a broad palette.

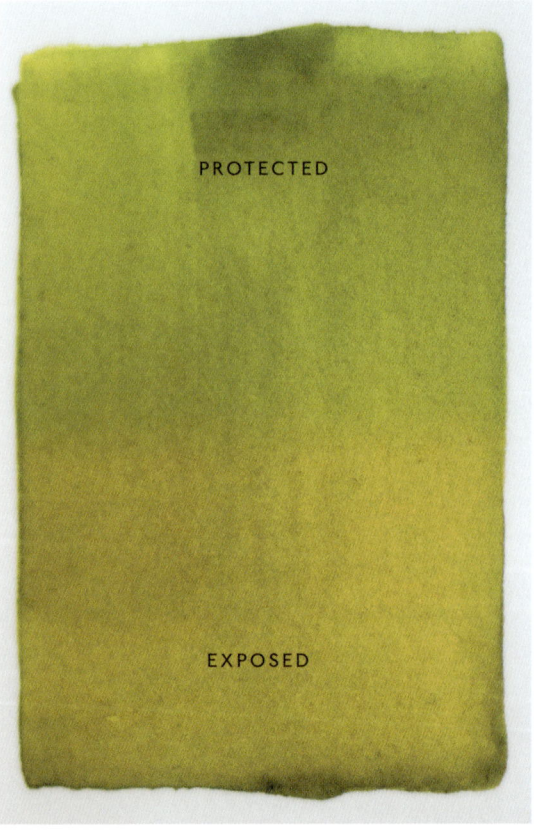

PROTECTED

EXPOSED

Effect of sunlight on sap green.

Lightfastness Testing

I approach lightfastness with a philosophical attitude. No materials are truly fixed or inert. They are substances that flow through space, across time, interacting with other substances.

Characterful, changeable botanical pigments signal their aliveness in the flow of matter. Engaging with them is a practice that values nuance over purity, relationship to nature more than control over nature, and cyclicality more than permanence.

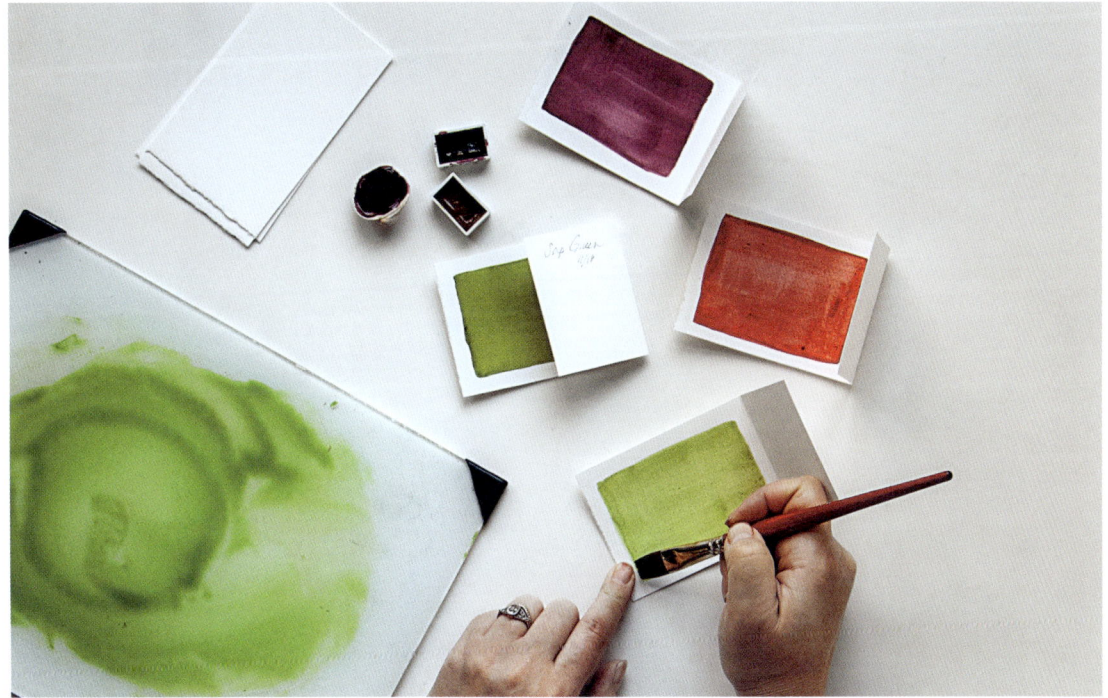

1. On a piece of heavy, acid-free paper, paint a swatch of washed lake pigment paste or paint. Place this on two-thirds of the paper's width, with one-third left blank. Fold the blank portion to cover half the swatch, and leave half exposed.

2. Tape the folded swatch in your sunniest window with the color facing outward, and label it with the pigment type and date. Leave it for 30 days in summer or 60 days in winter.

3. Unfold and compare the exposed pigment to the protected section. Any difference in color is the degree of fading you can expect with time and indoor light exposure.

The pH Scale

The manipulation of pH plays a massive role in both natural dyeing and laking. But what exactly are acids and bases, and how do they cause the chemical changes we observe in the studio/laboratory?

Water (H_2O) is a polarized molecule, bearing a partial positive charge on its hydrogen atoms, balanced by a partial negative charge on the oxygen atom. An important property of water is that it undergoes self-ionization to produce low concentrations of hydronium ions (H_3O^+) and of hydroxide ions (OH^-). In pure water the concentrations of hydroxide ions and hydronium ions are equal. When an ionic compound like alum dissolves in water, the balance is disrupted. If the concentration of hydronium ions becomes greater than hydroxide ions, the solution is said to be acidic. Conversely, a solution with a higher measure of hydroxide ions is said to be basic. The greater the superabundance, the stronger the acidity or basicity.

The ions in an acidic or basic solution make it rich with potentiality—it is charged. We measure this "potential of hydrogen" with the pH scale, running from 1 to 14. A pH value of 1 indicates a highly acidic solution (a large excess of hydronium ions over hydroxide ions); pH 7 is neutral (ions in equal concentration, as in pure water); pH 14 indicates a highly basic solution (a large excess of hydroxide ions). The pH scale is logarithmic, meaning that the step between each digit is a tenfold difference in the concentration of hydronium ions. A solution measuring pH 4 is 10 times more acidic than a solution measuring pH 5, and, likewise, 100 times more acidic than a solution measuring pH 6.

The laking process relies on pH manipulation to transform aluminum sulfate into aluminum hydroxide. Aluminum sulfate makes an acidic solution in water. When a base, such as sodium carbonate, is introduced it neutralizes the acid, leading to the formation of aluminum hydroxide ($Al(OH)_3$). Aluminum hydroxide is not soluble in neutral water, and it precipitates out of solution.

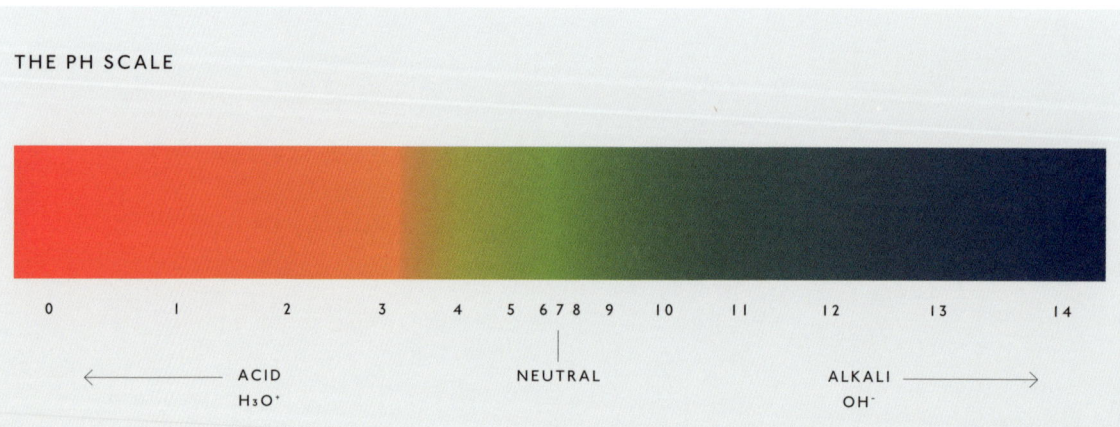

THE PH SCALE

0   1   2   3   4   5 6 7 8 9   10   11   12   13   14

←——— ACID   NEUTRAL   ALKALI ———→
H_3O^+ OH^-

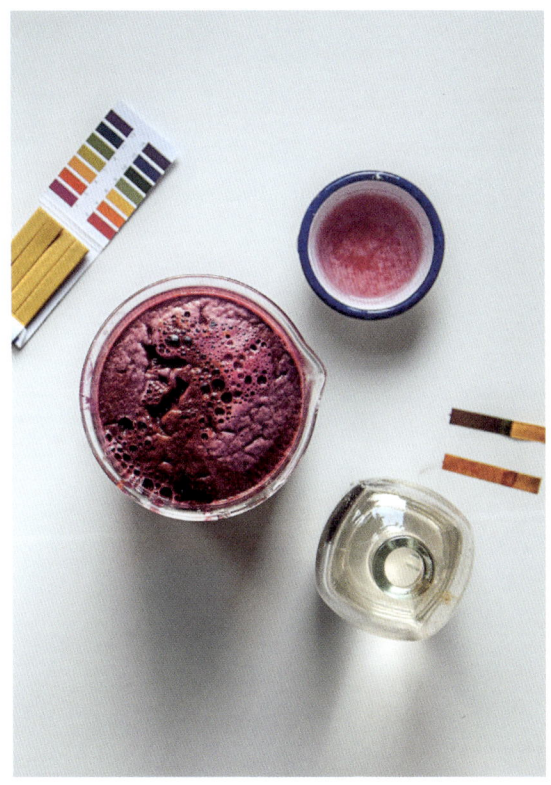

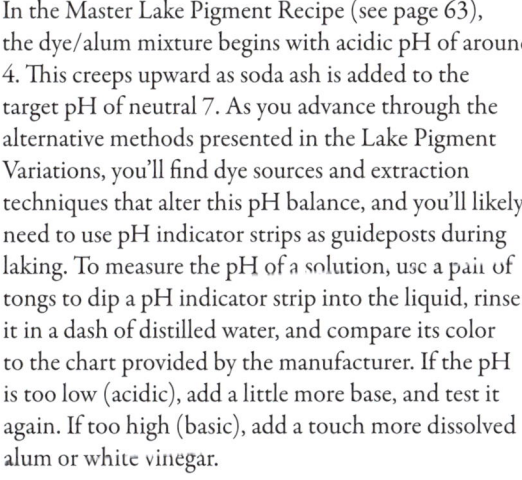

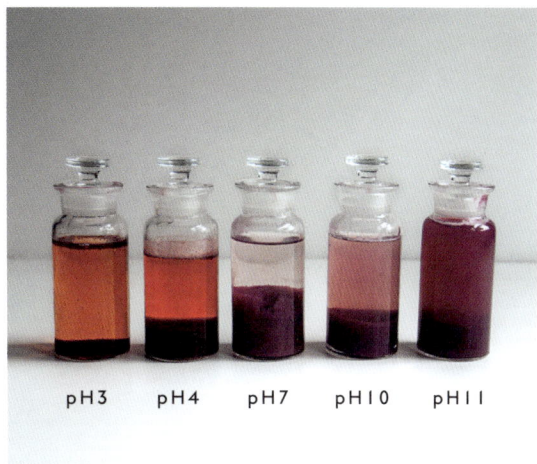

Effect of pH on Sappanwood Lake.

In the Master Lake Pigment Recipe (see page 63), the dye/alum mixture begins with acidic pH of around 4. This creeps upward as soda ash is added to the target pH of neutral 7. As you advance through the alternative methods presented in the Lake Pigment Variations, you'll find dye sources and extraction techniques that alter this pH balance, and you'll likely need to use pH indicator strips as guideposts during laking. To measure the pH of a solution, use a pair of tongs to dip a pH indicator strip into the liquid, rinse it in a dash of distilled water, and compare its color to the chart provided by the manufacturer. If the pH is too low (acidic), add a little more base, and test it again. If too high (basic), add a touch more dissolved alum or white vinegar.

Neutral pH is vital to the formation of lake pigments because aluminum possesses a peculiar characteristic called amphoterism, meaning it can react with both acids and bases. For our practical purposes in the studio, this means that aluminum can dissolve in either pH extreme. If the pH of a lake is too low, the aluminum hydroxide will not materialize. If excess alkali is added, the aluminum hydroxide will initially coalesce, but as the pH climbs too far above neutral the pigment will dissolve back into solution. Tinkering with pH is a strong temptation for the natural dyer, because many plant colorants express a rainbow of hues along the pH spectrum. For example, the anthocyanins of purple flowers like hibiscus and hollyhock shift from green through purple and pink in reaction to pH. For the stability of a lake pigment, fight the temptation of pH color modification and keep it between 6 and 8. As the pH veers away from this safe zone, the pigment yield diminishes to the point of nonexistence.

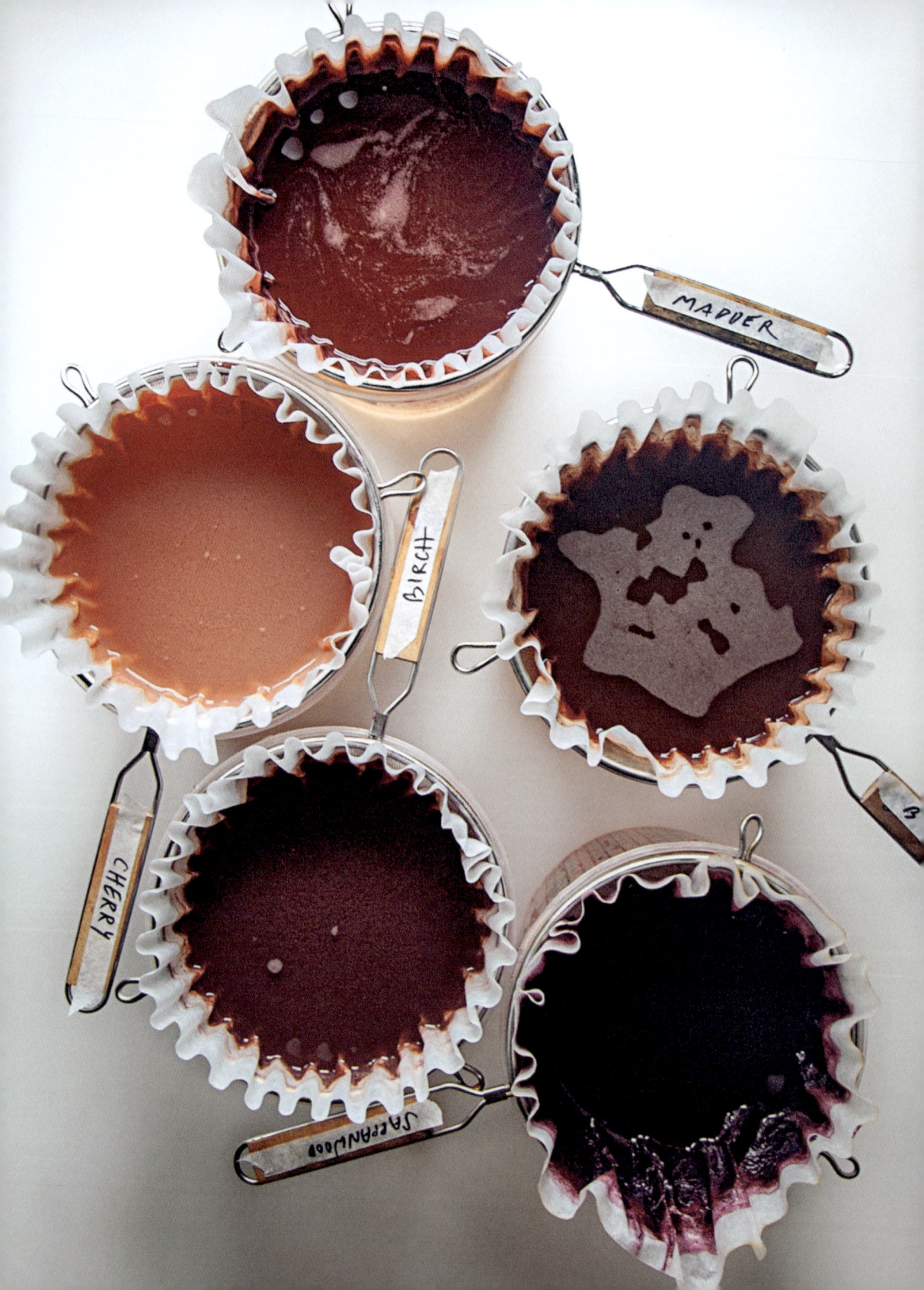

Lake Pigment Variations

The Master Lake Pigment Recipe (page 63) provides a foundational understanding of the laking process and might be the only method you use to create a wide array of botanical pigments; however, each of the following variations explore an alternative extraction method or modification, along with some of the plants well suited to it. These demonstrate the expressive power of every plant source to produce a nuanced spectrum and maximal color yield. Inspired by artisans' recipes from more than two centuries ago, as well as ongoing research and reclamation of natural color heritage, I hope that these recipes offer a springboard for practical experimentation and deeper engagement.

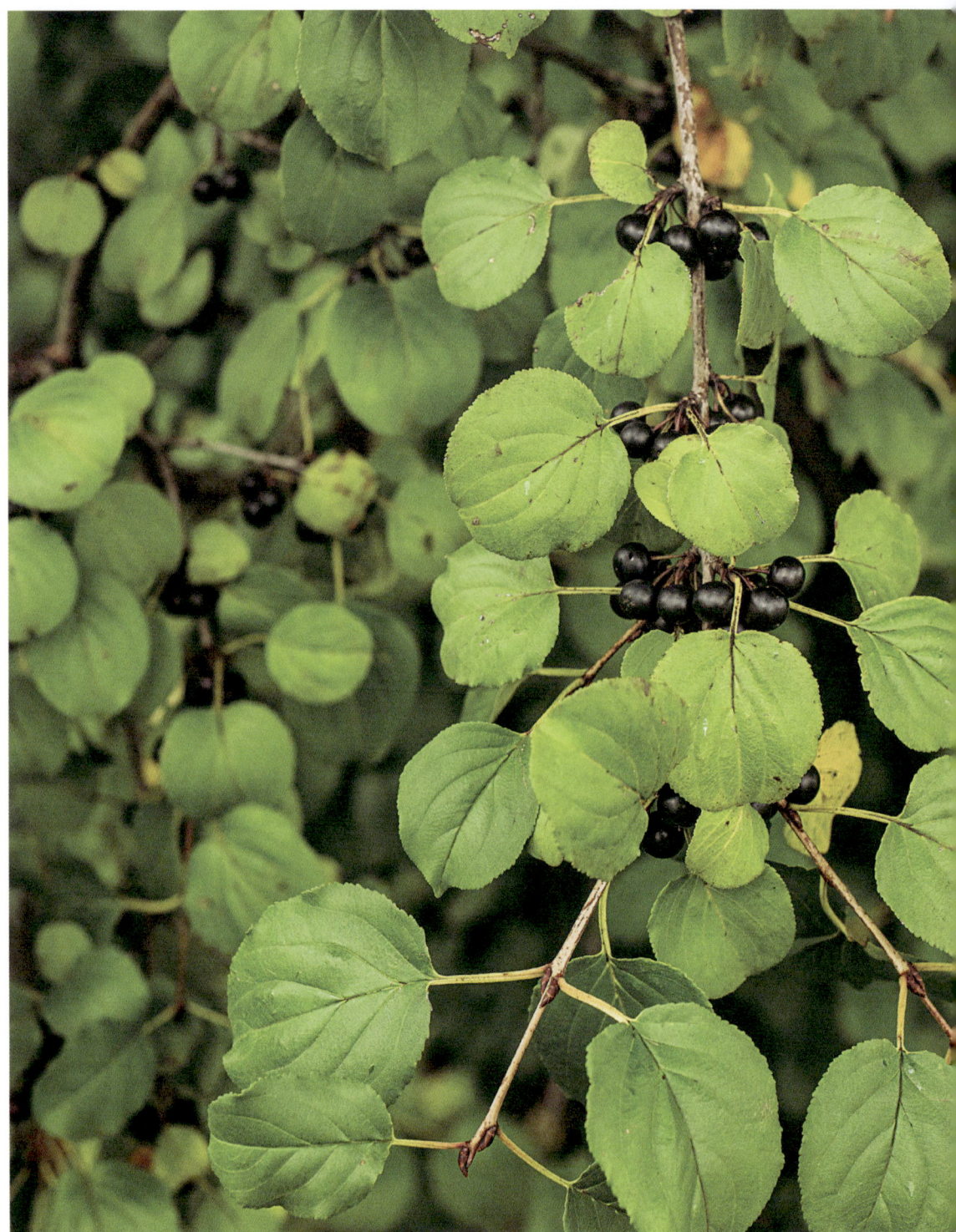

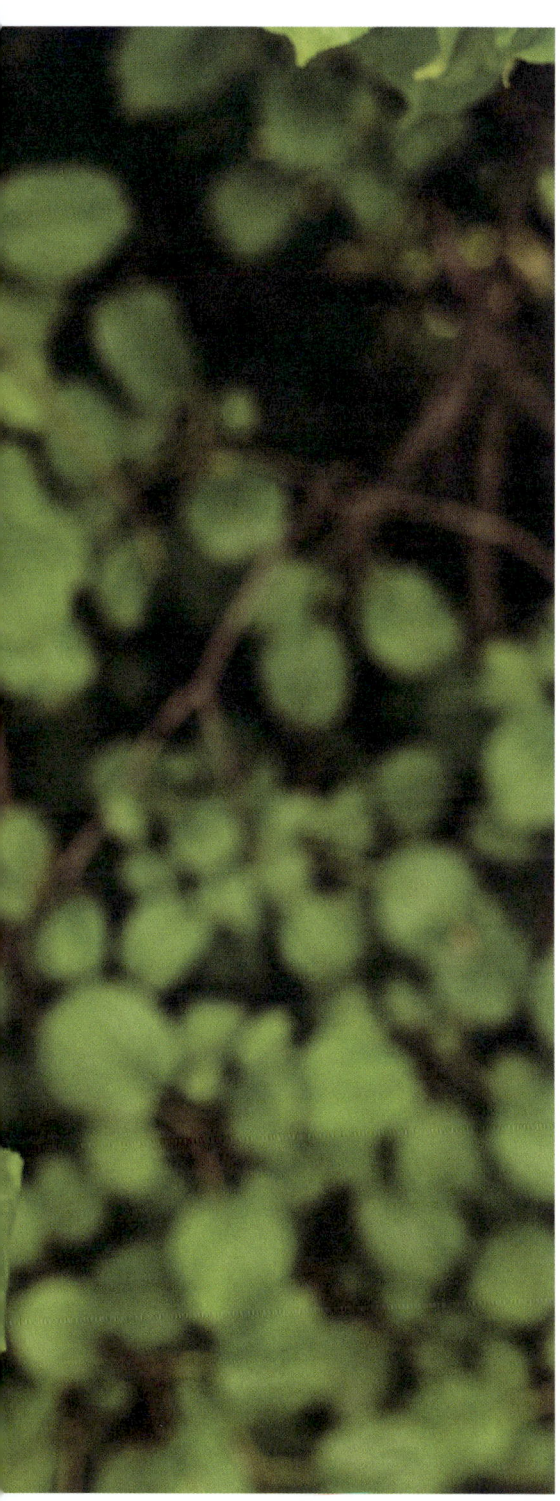

Buckthorn

This lake pigment variation includes an optional waylay with sap green, the sticky, concentrated juice of buckthorn berries. It has been used as water-based ink since at least the medieval era. The jewel-like berry juice is fortified with alum and visibly demonstrates the importance of the mordant. At the instant alum enters the dye, the color immediately and strikingly stabilizes and vivifies to a vegetal yellow-green. The dye can be further transformed into Buckthorn Lake with the addition of an alkali. These two metamorphoses provide a case study of our essential ingredients.

Common buckthorn (*Rhamnus cathartica*) is a vigorous, spiny shrub that thrives along woodland margins and roadsides. In the spring, it produces small green berries, which mature to a plump purple-black in autumn. The berries provide an array of colors across the growing season, starting with yellow and maturing through shades of green. Sap Green requires ripe berries, which can be foraged or sourced from a dye vendor. (Note that commercially available green buckthorn berries are underripe. The berries are green, but they produce yellow dye.) Dye colors also can be extracted from the leaves and inner bark. Note that all parts of the buckthorn plant contain purgative and laxative compounds, so handle and store them with care. Work in a ventilated area or outdoors during dye extraction.

In the United States, common buckthorn is so robust that many states legally designate it an invasive species. Before foraging, check its status in your area. For more information about overabundant plant species, see Foraging Ethics (page 22).

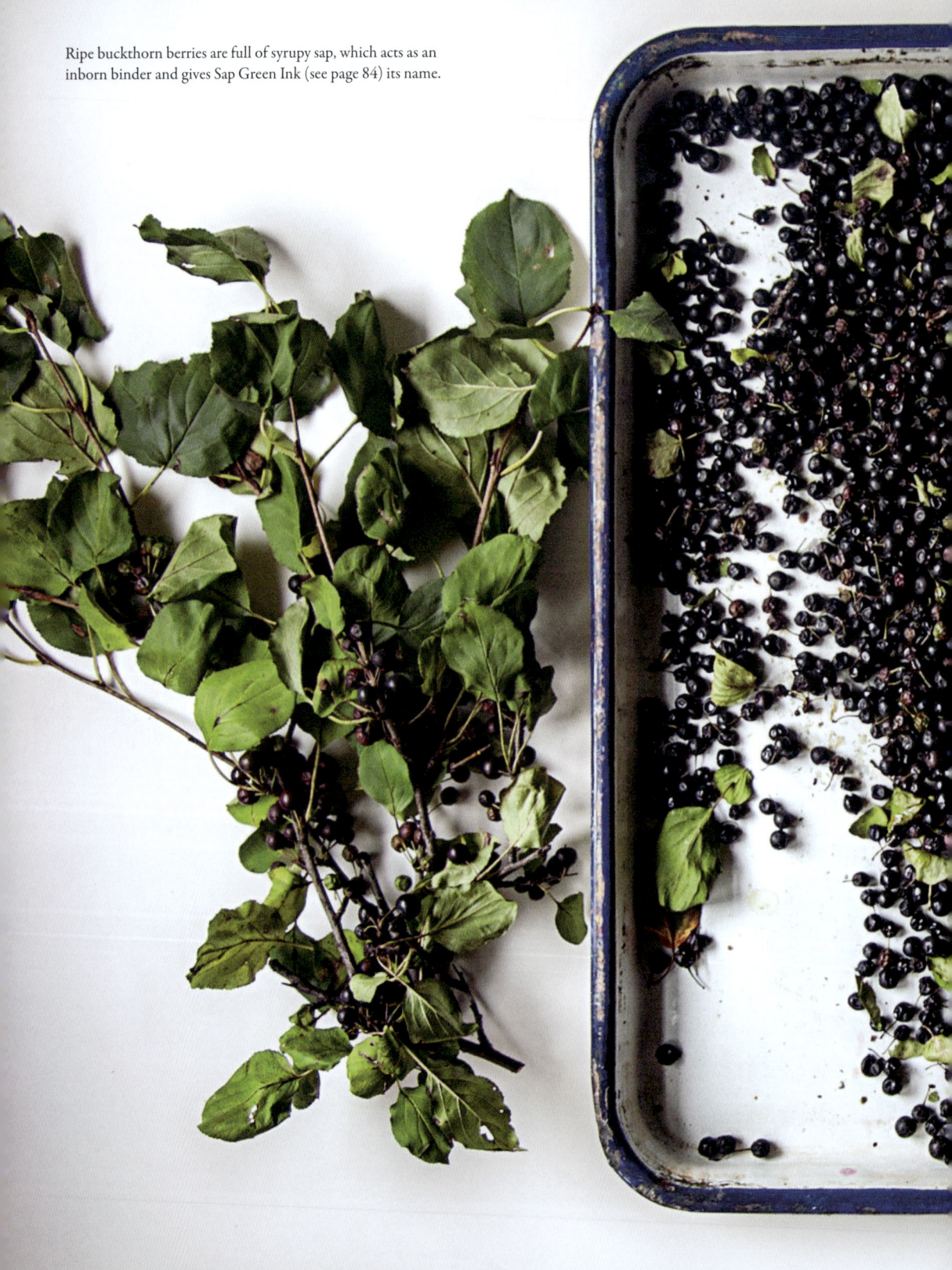

Ripe buckthorn berries are full of syrupy sap, which acts as an inborn binder and gives Sap Green Ink (see page 84) its name.

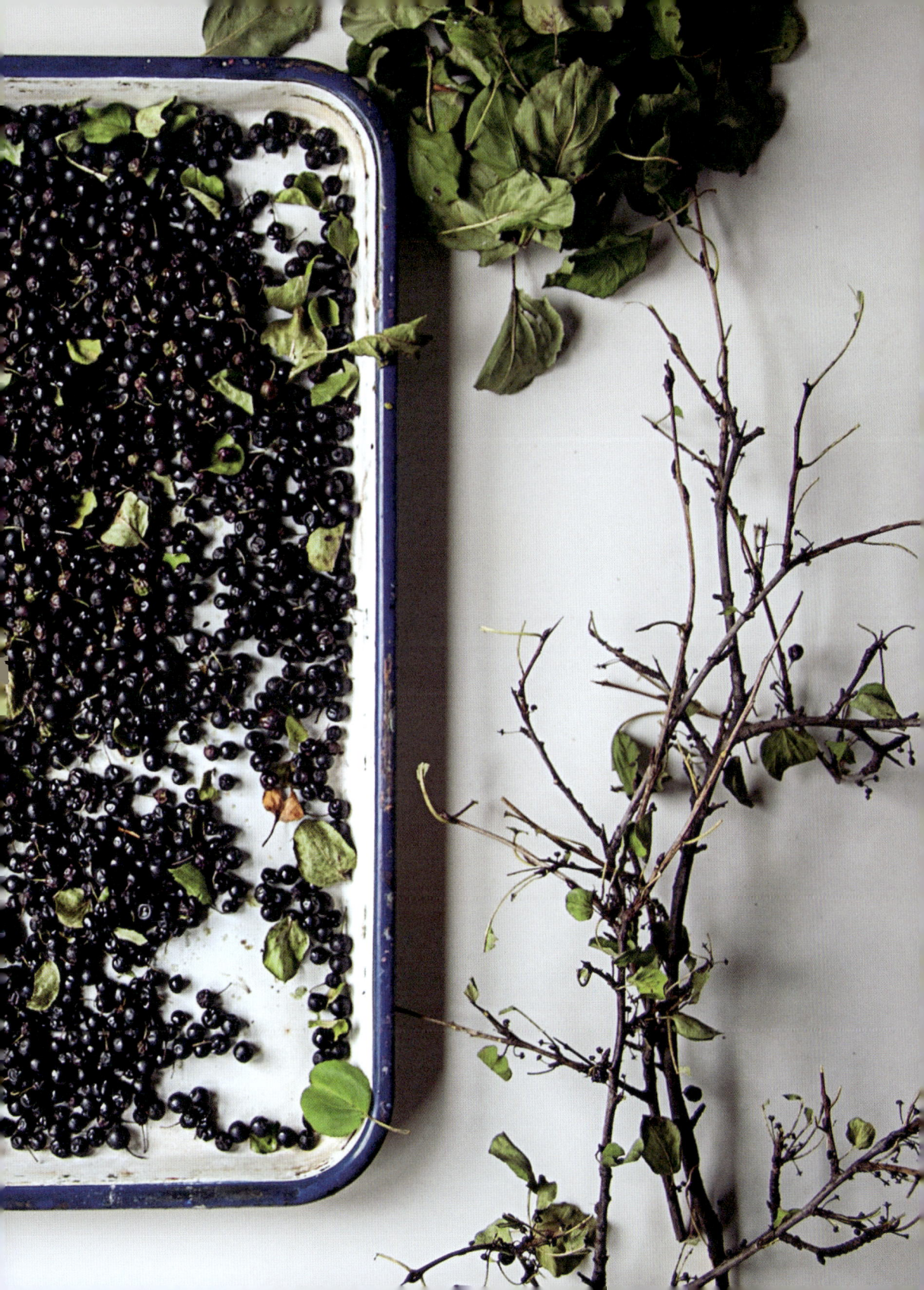

SAP GREEN INK

Carefully reducing the liquid of ripe buckthorn berries over low heat thickens it to a viscous paint-like consistency without any additional ingredients. Before the advent of aluminum paint tubes, painters stored the tarry substance in a pig's bladder, giving it the alternate name "bladder green." A seashell, nutshell, or small jar works just as well.

MATERIALS

· 20g dried ripe buckthorn berries (or 100g fresh)

· 3g alum

· Distilled water

· Double boiler

· Small storage jar or seashell

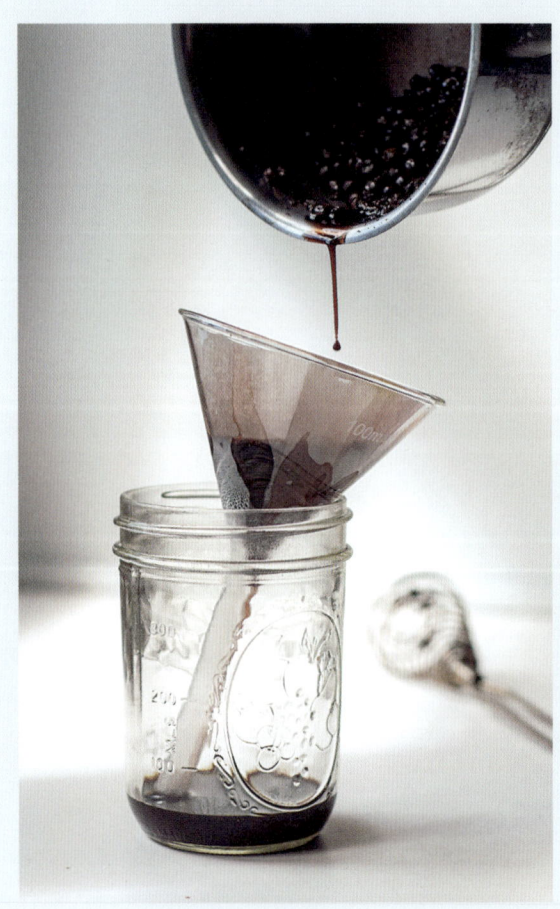

METHOD

1. In a small pot, cover ripe buckthorn berries with 750ml distilled water and simmer 30 to 60 minutes. After a few minutes of softening in the hot water, use a masher or spoon to crush the berries and release their juice. Cover the pot or top up with additional water as needed to prevent it from boiling dry.

2. Filter the liquid through a coffee filter, then discard the spent dyestuff. The liquid is your dye, rich with the berries' soluble colorants. It should be a translucent jewel-like russet color, but this can vary considerably with the degree of ripeness.

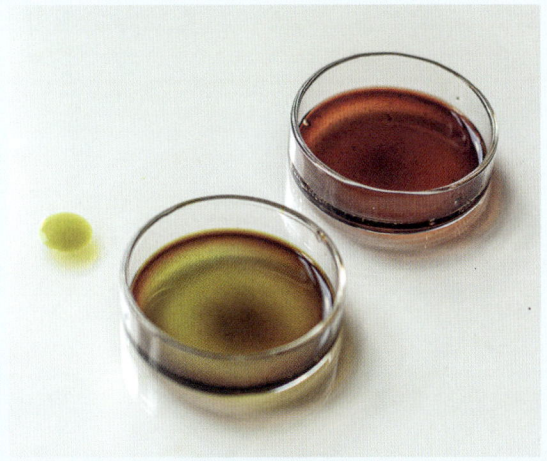

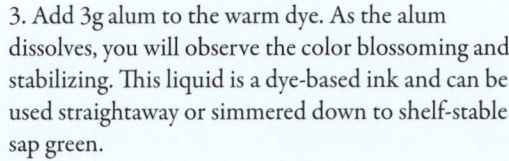

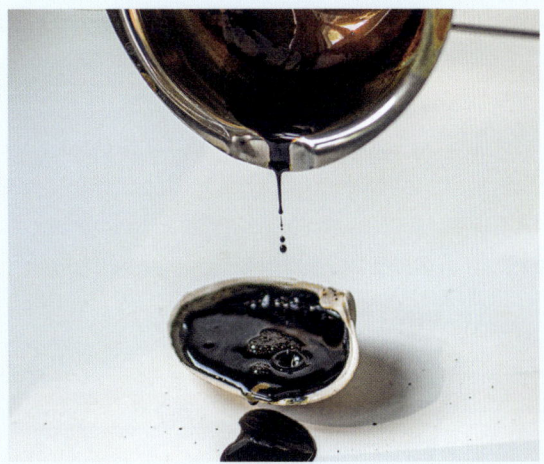

3. Add 3g alum to the warm dye. As the alum dissolves, you will observe the color blossoming and stabilizing. This liquid is a dye-based ink and can be used straightaway or simmered down to shelf-stable sap green.

4. Transfer the liquid to the top of a double boiler with a water bath below. This will moderate the heat and prevent the ink from scorching. Return to medium-high heat and boil uncovered until the green ink is concentrated. Top up the water bath as needed. When the ink has thickened to a tacky, syrupy consistency, remove it from heat. Before it is too thick to maneuver, transfer to a seashell or storage jar.

5. Store at room temperature, uncovered. Sap green will gradually dry to a stiff and tarlike consistency, but it remains easy to rewet. To use, moisten the surface with a wet paintbrush, as you would a watercolor pan.

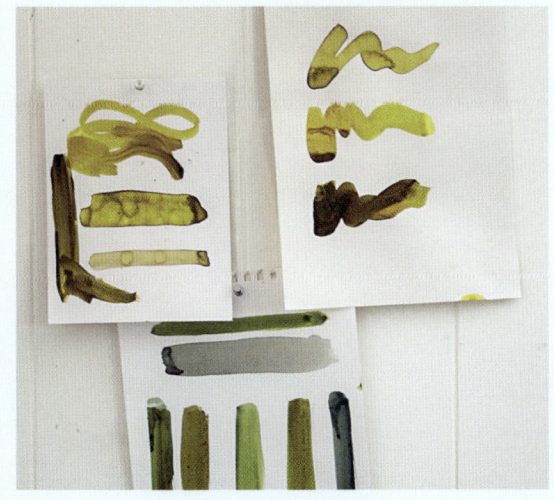

The component parts of a dye-based ink like sap green all remain water-soluble. Even if dried to a crisp and ground to dust, the dye, mordant, and binder will all drift apart into translucent solution when rehydrated. This is the fundamental distinction between dyes and pigments: Dyes dissolve, and pigments do not.

BUCKTHORN LAKE PIGMENT

This green lake pigment is made with the dye extracted from ripe berries. Keep in mind that underripe or overripe buckthorn can be used in the same way to produce different shades. The most lightfast colors come from the early season, before the berries ripen.

MATERIALS

· 10g buckthorn berries, dry

· 10g alum

· 5g soda ash (sodium carbonate)

· Distilled water, for washing pigment

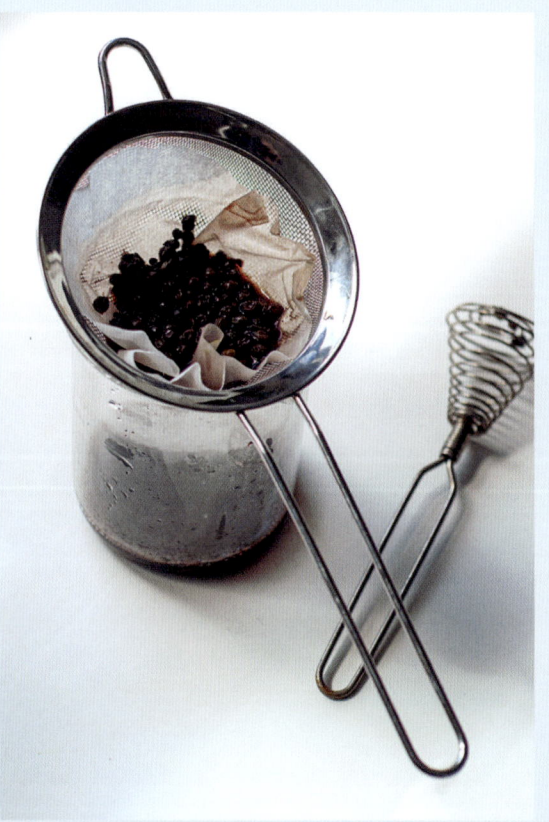

METHOD

1. Cover buckthorn berries with approximately 750ml water. Simmer for 30 to 60 minutes to extract the dye. Cover the pot or top up with additional water as needed to prevent it from boiling dry. Filter the dye, and discard the spent dyestuff.

2. Transfer the dye to a 1-liter container, leaving ample headspace for the bubbly neutralization reaction. Add hot water to equal 500ml if needed. Add 10g alum; stir to dissolve.

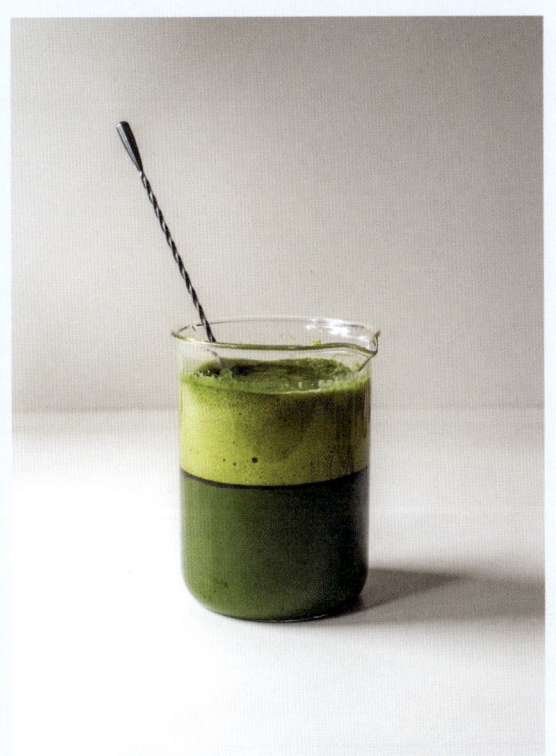

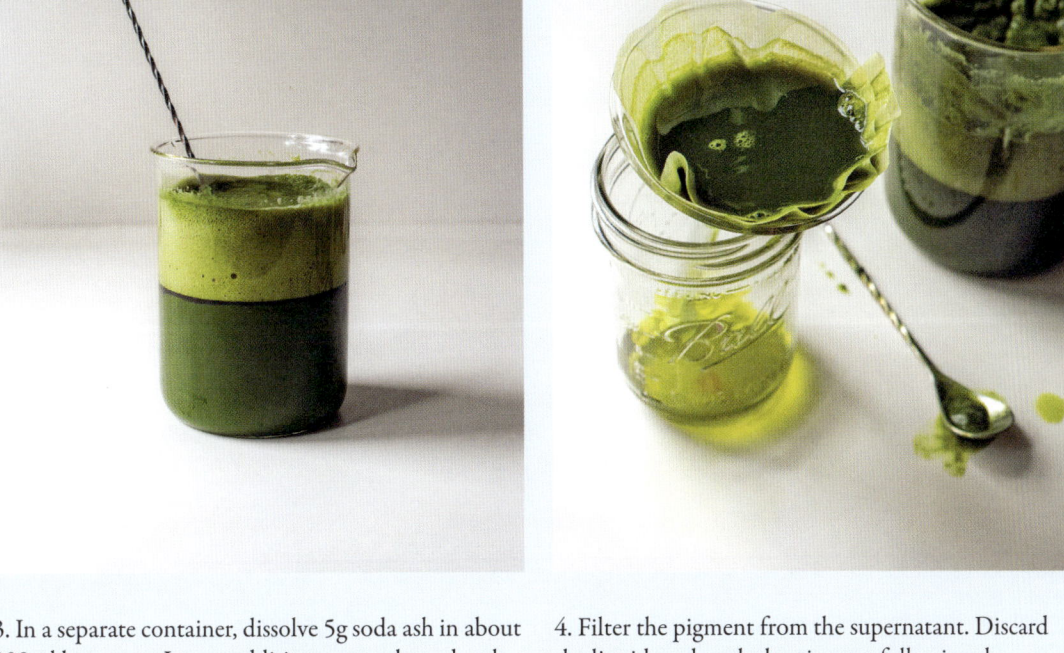

3. In a separate container, dissolve 5g soda ash in about 200ml hot water. In two additions, pour the soda ash solution into the buckthorn liquid, stirring slowly between each.

4. Filter the pigment from the supernatant. Discard the liquid, and wash the pigment following the instructions in the Master Lake Pigment recipe (page 63), starting at step 14.

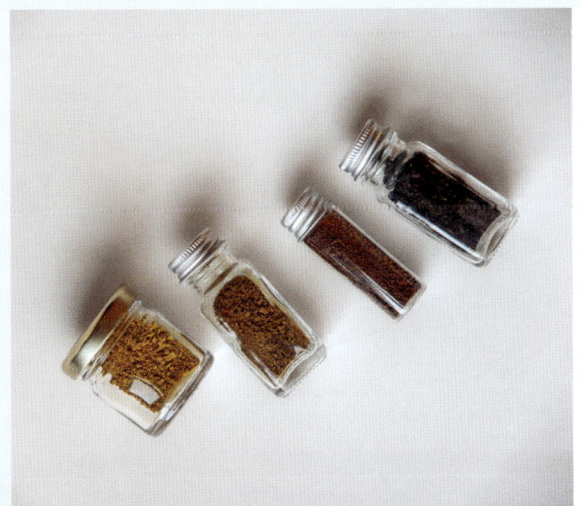

OTHER PLANTS FOR SAP GREEN INK & LAKE PIGMENT

There are many varieties of buckthorn suitable for ink and lake, including *Rhamnus infectoria*, *R. saxatilis*, *R. tinctoria*, and *R. alaternus*. *Frangula purshiana* (cascara) and *Frangula alnus* (alder buckthorn) can give similar results. *Hippophae* spp. (sea buckthorns) are not a substitute.

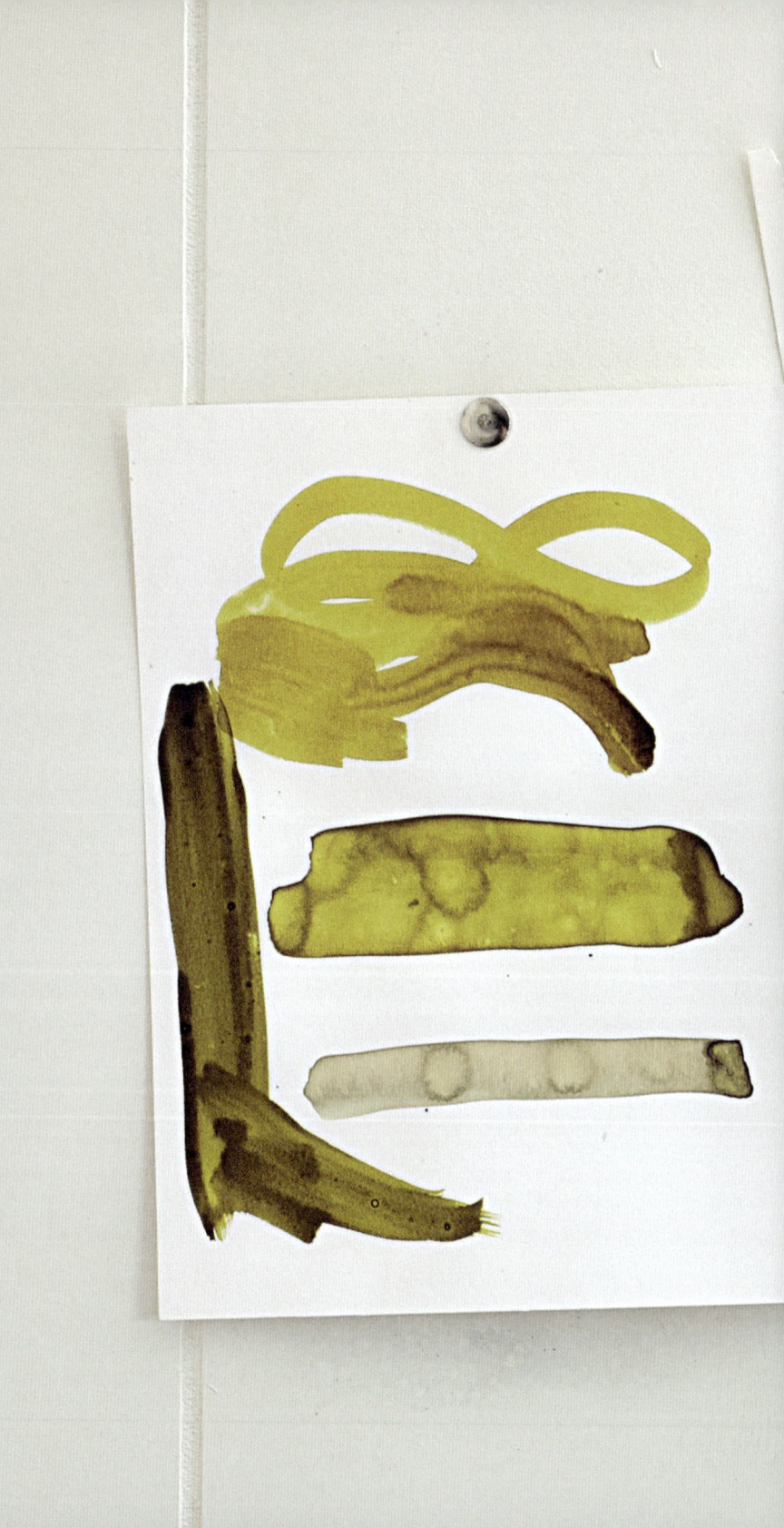

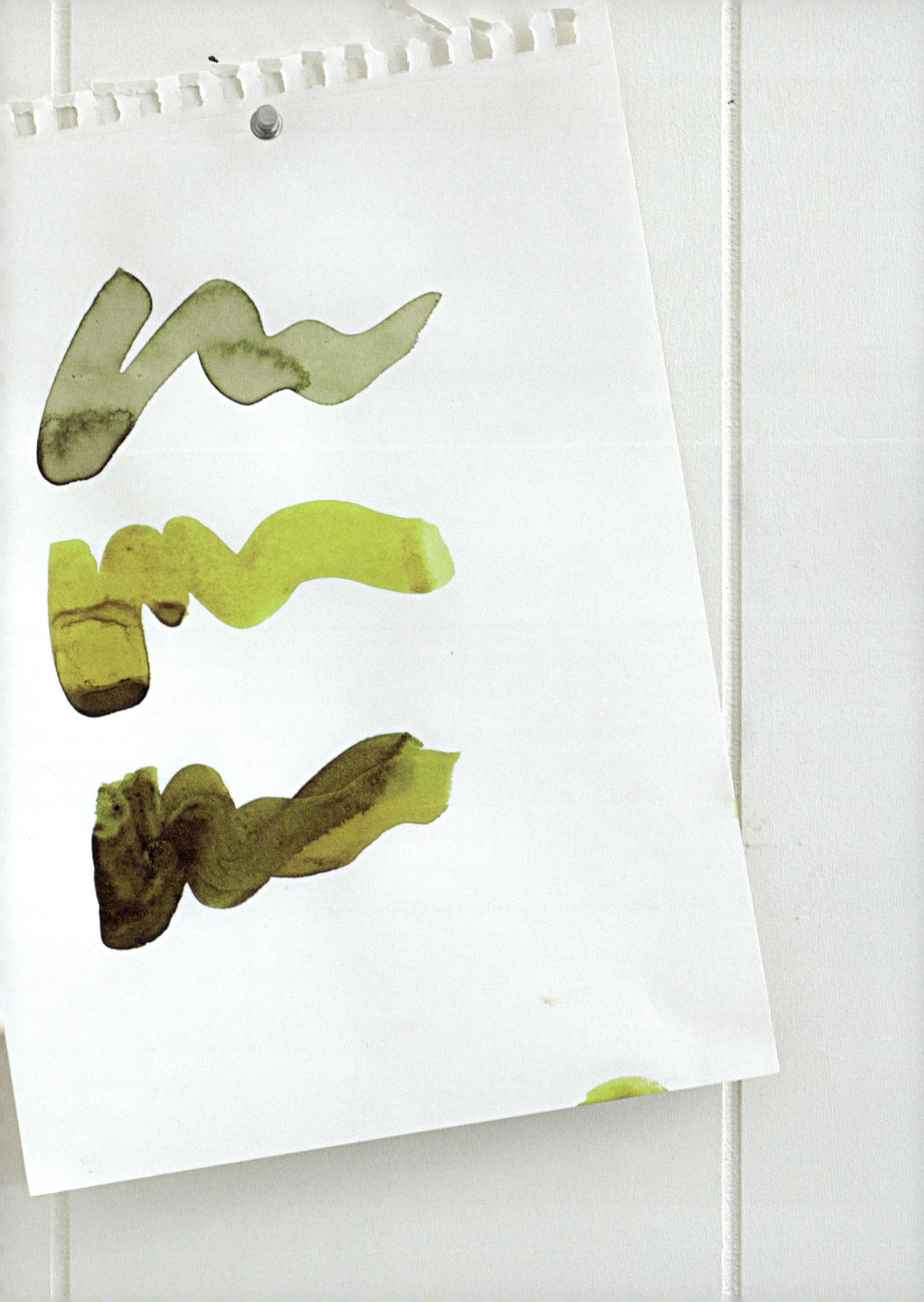

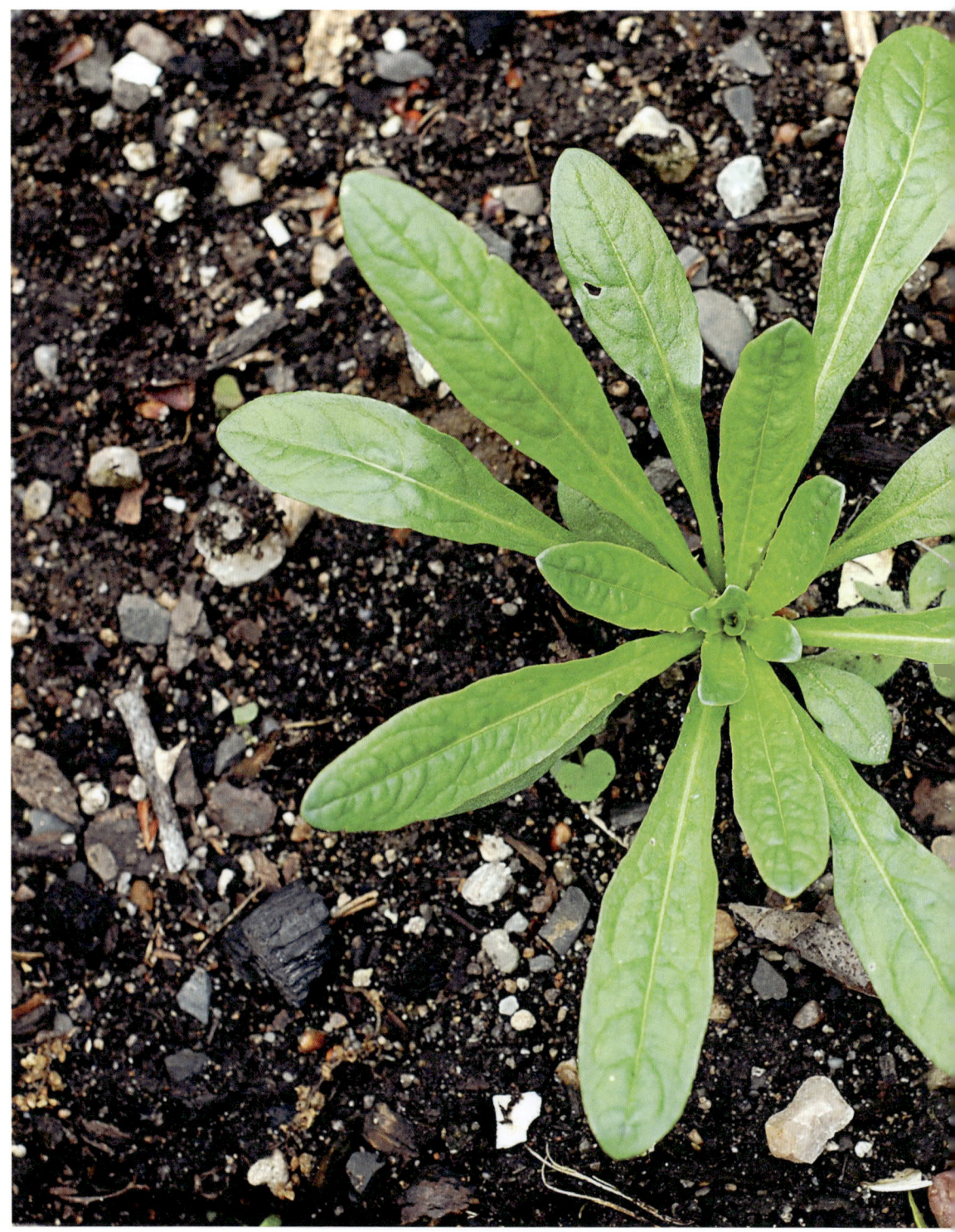

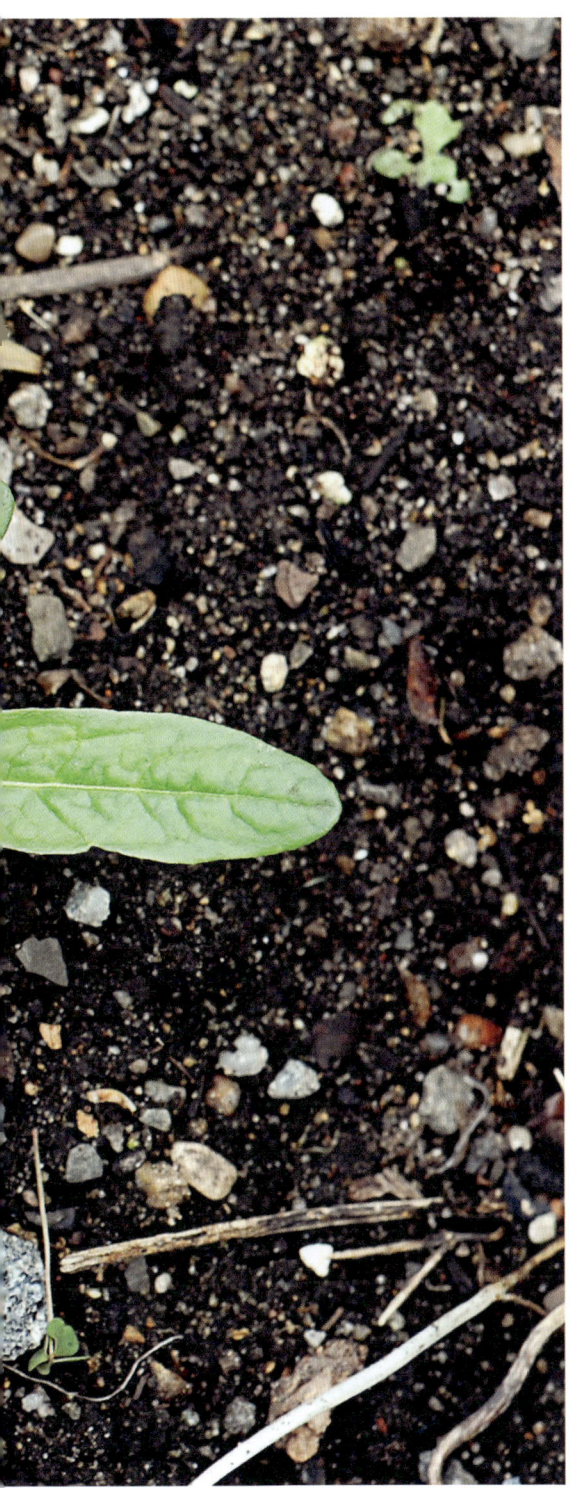

Weld

Weld (*Reseda luteola*) is the queen of yellows: It's a clear lemon for mixing both warm and cool tones, and it features the greatest lightfastness of the yellow dyes. It's easy to grow from seed in full sun on well-drained ground. In its first year, the plant produces a modest rosette of leaves, which can be individually plucked for dye extraction. However, the main harvest comes in weld's second year, when a tall spire of blossoms bolts upward. This occurs in mid-spring along with the flowering of other biennials like foxglove and columbine, which have a head start over the annuals just starting to show their sleepy heads.

I leave a couple of vigorous weld plants to set seed and harvest all the rest just as the seed pods start to ripen. In this way, weld happily shares a garden bed with other dye plants, ceding the growing space just when annuals start to need it. After growing weld once, self-sown seedlings will likely sprout on their own. If they need to be moved around the garden, do so during their first autumn while they're still quite small. The flowers and leaves contain the most dye, and the stalk itself can be composted.

WELD LAKE PIGMENT

Weld has a funkily pungent scent in the dye pot, and the scent strengthens over time. Process this pigment expeditiously and wash it thoroughly, before it becomes aromatic.

MATERIALS

· 12g dried weld

· 0.5 to 1ml 10 percent ferrous sulfate solution, or homemade iron solution (optional)

· 10g alum

· 5g soda ash

· 2 pinches of unflavored gelatin powder, equivalent to about one-tenth of a gold-strength gelatin leaf (optional)

· Distilled water, for washing pigment

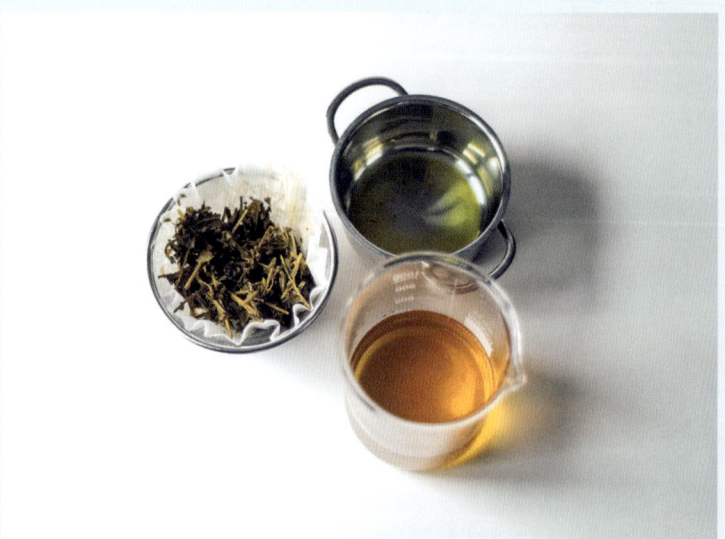

METHOD

1. Cover weld with approximately 750ml water. Simmer for 30 to 60 minutes to extract dye. Filter the dye and compost or discard the spent dyestuff.

2. Optional: For Color Modification with Iron, turn to page 94.

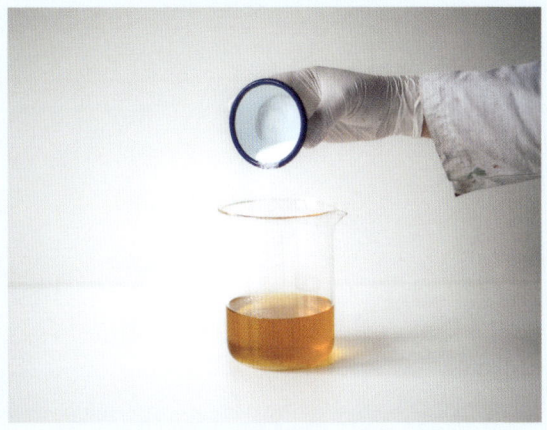

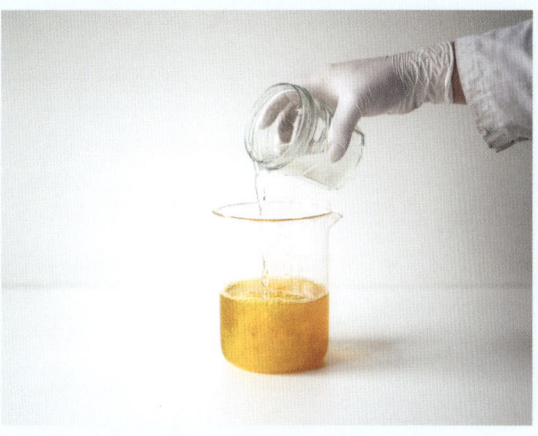

3. Transfer the dye to a 1-liter container. Add hot water to equal 500ml, if needed. Add 10g alum; stir to dissolve.

4. In a separate container, dissolve 5g soda ash in about 200ml hot water. Pour roughly half the soda ash solution into the dye mixture and gently stir. Repeat with the remaining soda ash.

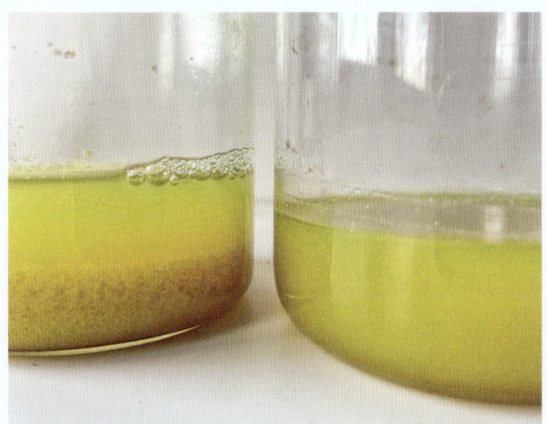

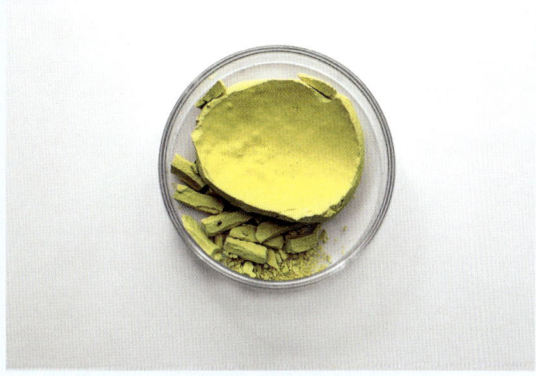

5. Occasionally, a lake pigment materializes in very minute particles that are exceptionally slow to settle and filter. I have experienced this with weld lakes, in particular. To speed up processing, you might like to flocculate your pigment, causing the particles to clump together in larger groups. Sprinkle two pinches of gelatin powder over a small dish of steaming hot water. Rest for five minutes to bloom. Then pour into your lake pigment mixture. After 10 minutes, you should see the mass of particulate settle out. Decant off the liquid above and filter the pigment.

6. Filter the pigment from the supernatant. Discard the liquid and wash the pigment following the instructions in the Master Lake Pigment recipe on page 63, starting at step 14.

COLOR MODIFICATION WITH IRON

The sunny yellow of a lake made from weld is also called "arzica." To modify the color with an iron mordant, turning it from lemon yellow to olive green, follow the simple steps below. All mordant dyes react to the saddening influence of iron, which darkens and neutralizes their vibrancy. Iron can be added to any other lake following this process. See pages 37–39 for more information on iron and instructions for homemade iron solution. Remember: Whenever you use iron, take care to scrub all your tools thoroughly to prevent any residue from impacting subsequent pigments.

METHOD

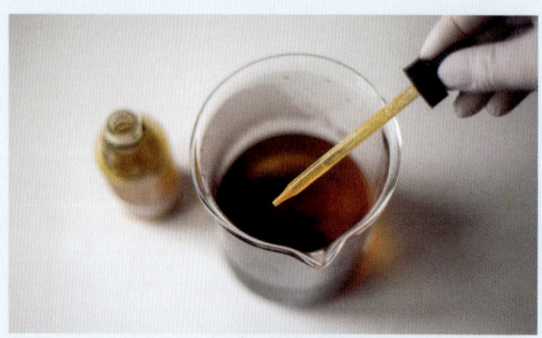

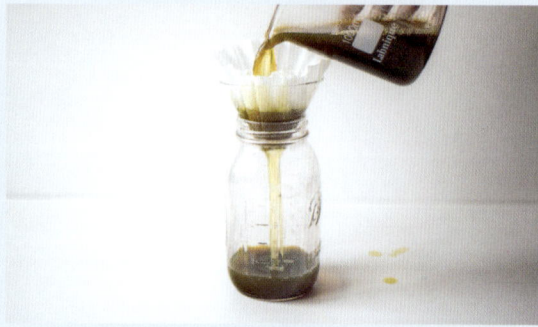

1. Add 0.5–1ml (5–10 drops) of iron solution (page 38 or 39) gradually until you observe a color change from lemon yellow to olive green.

2. Filter again to remove any oxidized iron specks. Return to the Weld Lake Pigment recipe on page 93, picking up at Step 3.

OTHER PLANTS FOR MODIFICATION WITH IRON

All mordant dyes are susceptible to iron's saddening effect. Other yellow-bearing plants—including marigold, goldenrod, yarrow, dyer's coreopsis, black-eyed Susan, and golden marguerite—likewise give shades of olive and sage with iron modification. Tannin-rich plants that tend toward dull browns can benefit from iron modification to produce deep tonal grays.

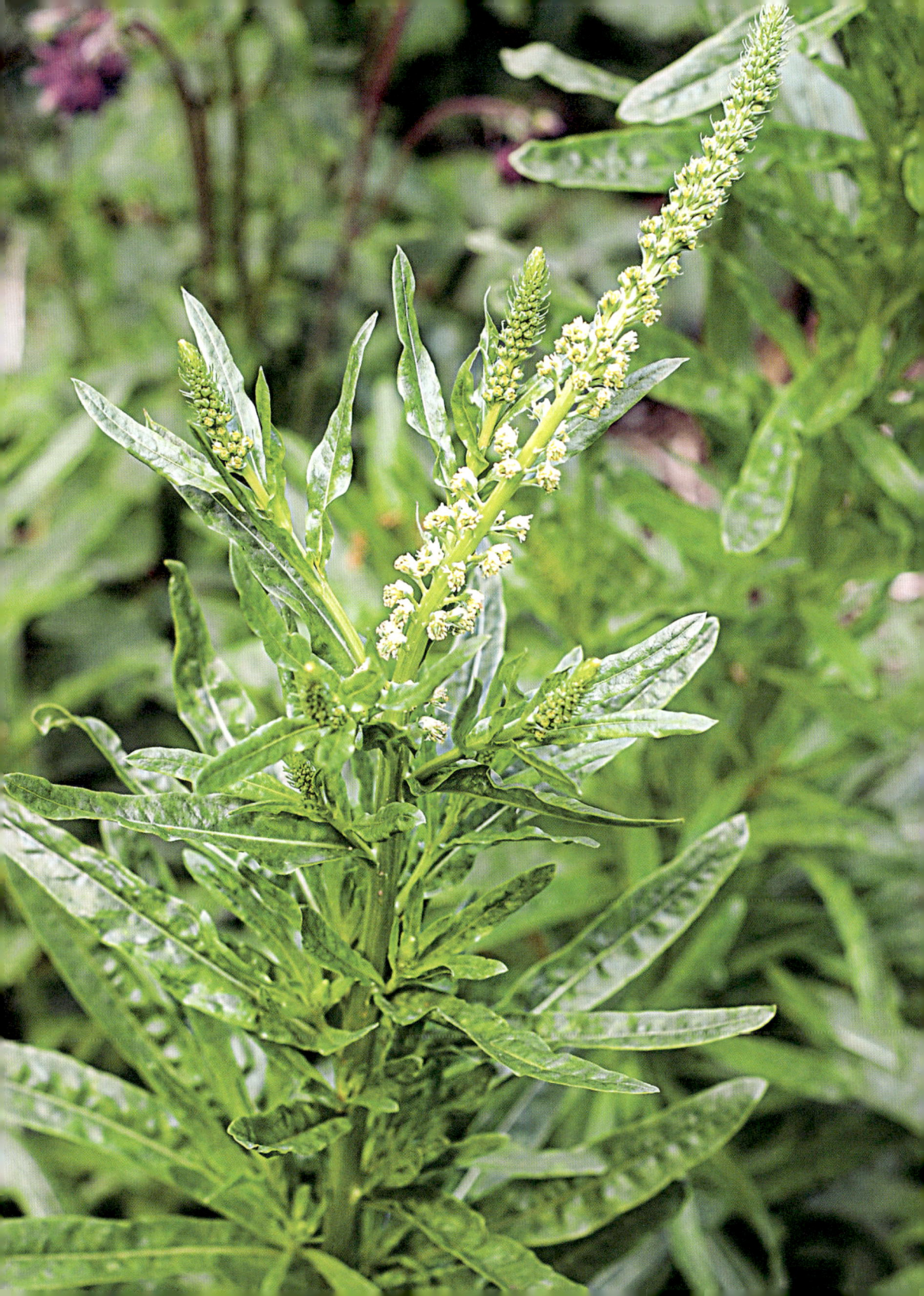

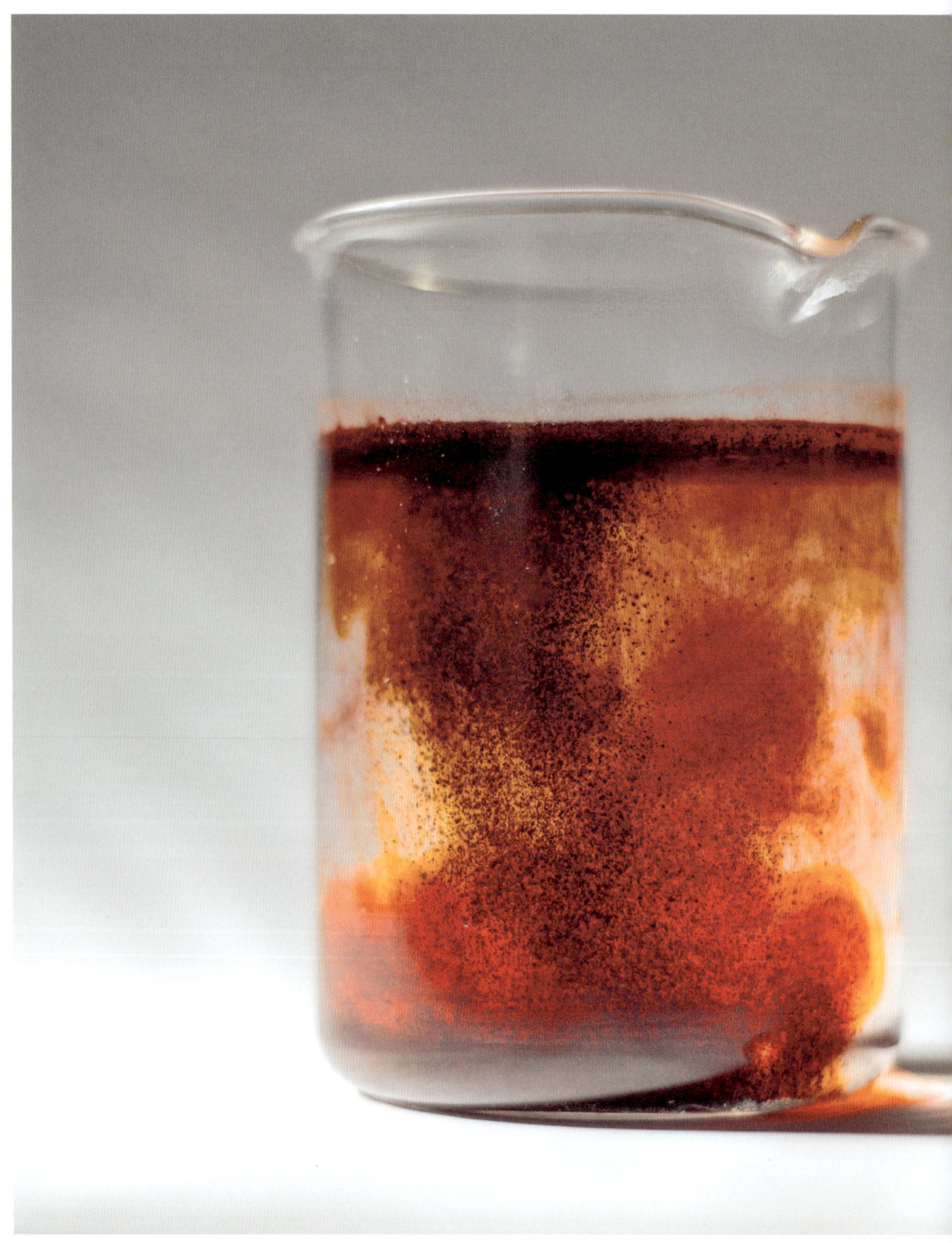

Lac Extract

Lac is not a botanical dye source, but it's a rather important insect dye from an historical and color perspective. Lac extract comes from the sticky secretion of the lac insect (*Kerria lacca*), cultivated in south and southeast Asia. These scale insects live on the tender branches of trees, protected in colonies by the resinous shell coating they produce. This material, stick lac, is purified to make shellac as well as lac dye. Lac is a renowned dye due to the rich hue and strong lightfastness of its primary colorant: laccaic acids. The beautiful shades of garnet and plum, produced from the relationship between lac insects and their host plants, demand thoughtful and respectful use.

Dye extract powders are concentrated pre-extracted dyes that simply dissolve in hot water like vegetable bouillon. Just as in cooking soup, the extract powder is convenient and timesaving; however, more nuance is possible when making an extraction from scratch. That said, resinous dyes like lac and cutch are primarily available in powder form because their color is impractical to extract at home.

OTHER DYE EXTRACTS FOR LAKING

Powdered or liquid dye extracts of any mordant dyes can be transformed into lake pigments.

LAC LAKE PIGMENT FROM EXTRACT

Lac dye produces velvety plum-colored pigments, and it also is the likely progenitor of the term "lake pigment." Lac comes from the Hindi *lakh* for "one hundred thousand," but, coincidentally, *lac* is also the French term for a lake.

MATERIALS

· 6g lac extract powder

· 10g alum

· 5g soda ash

· Distilled water for washing pigment

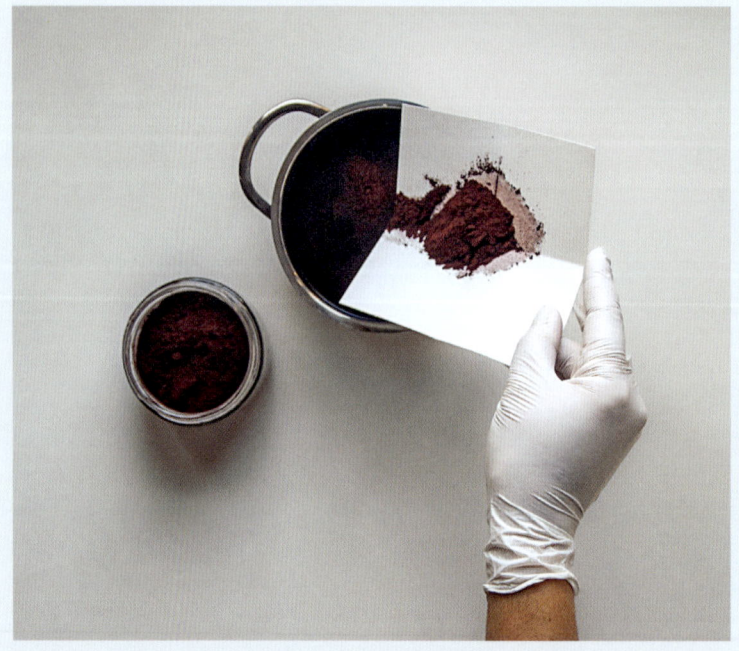

METHOD

1. In a saucepan, bring about 500ml water to boil. Remove from heat. Stir in lac extract powder. Rest for 10 to 20 minutes, stirring occasionally.

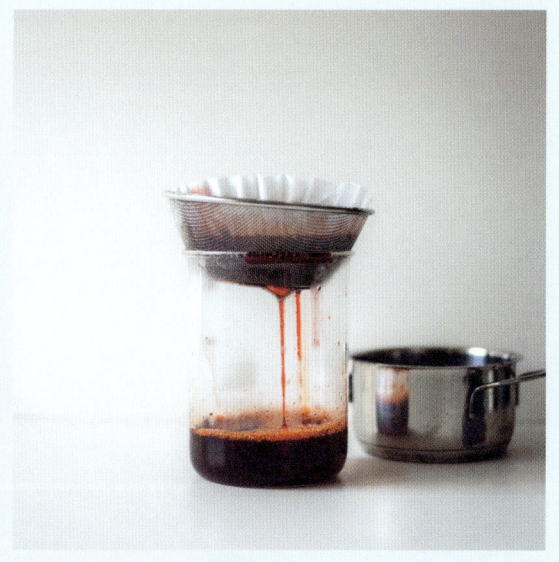

2. Filter the dye to remove any resin or impurities.

3. Transfer the dye to a 1-liter container. Add hot water to equal 500ml if needed. Add 10g alum, and stir to dissolve.

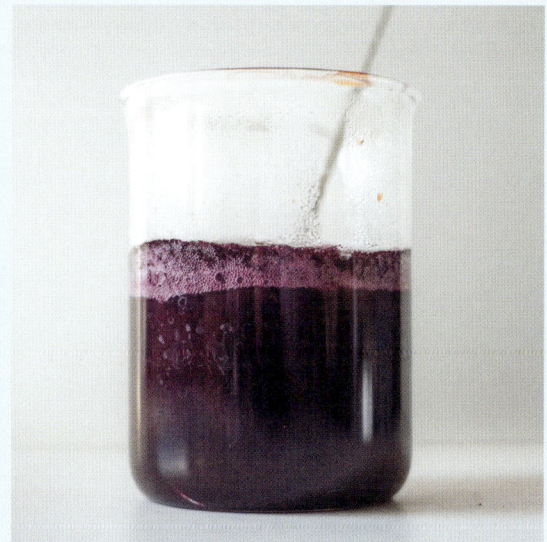

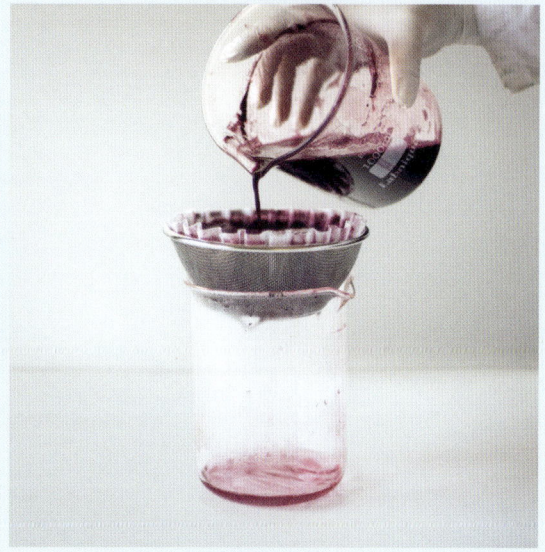

4. In a separate container, dissolve 5g soda ash in about 200ml hot water. Pour roughly half the soda ash solution into the dye mixture; gently stir. Repeat with the remaining soda ash.

5. Filter the pigment from the supernatant. Discard the liquid, and wash the pigment following the instructions in the Master Lake Pigment recipe on page 63, starting at step 14.

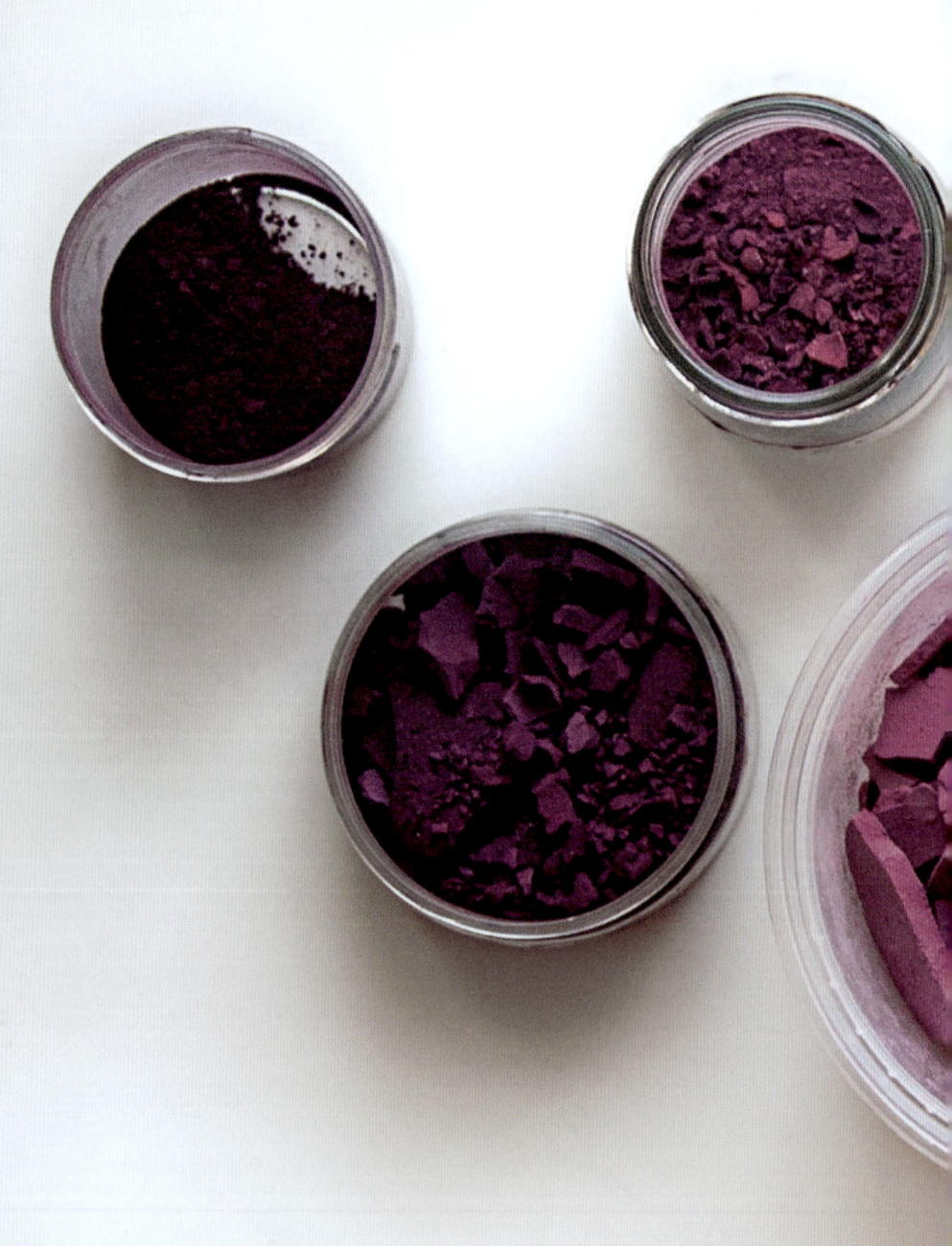

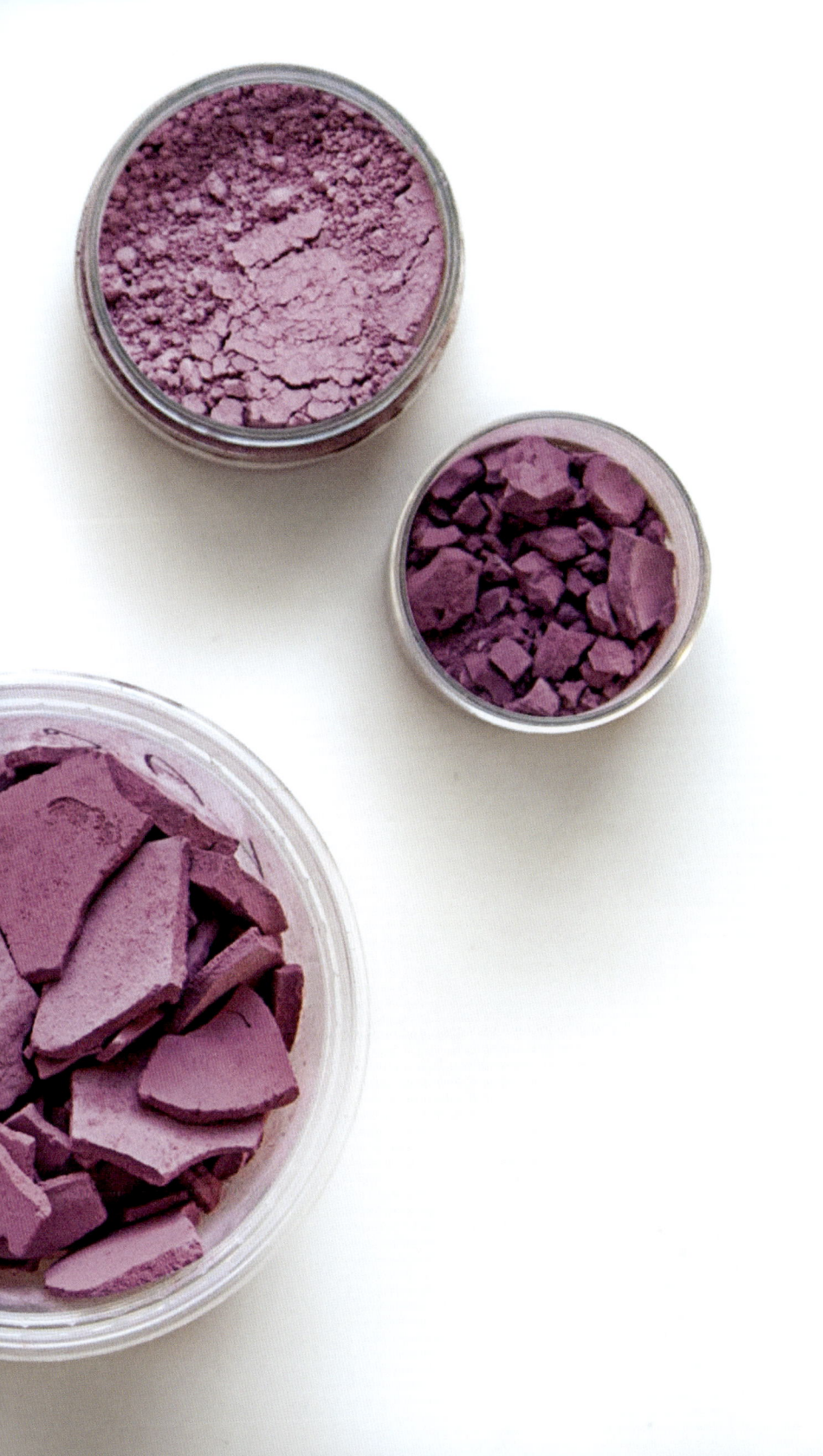

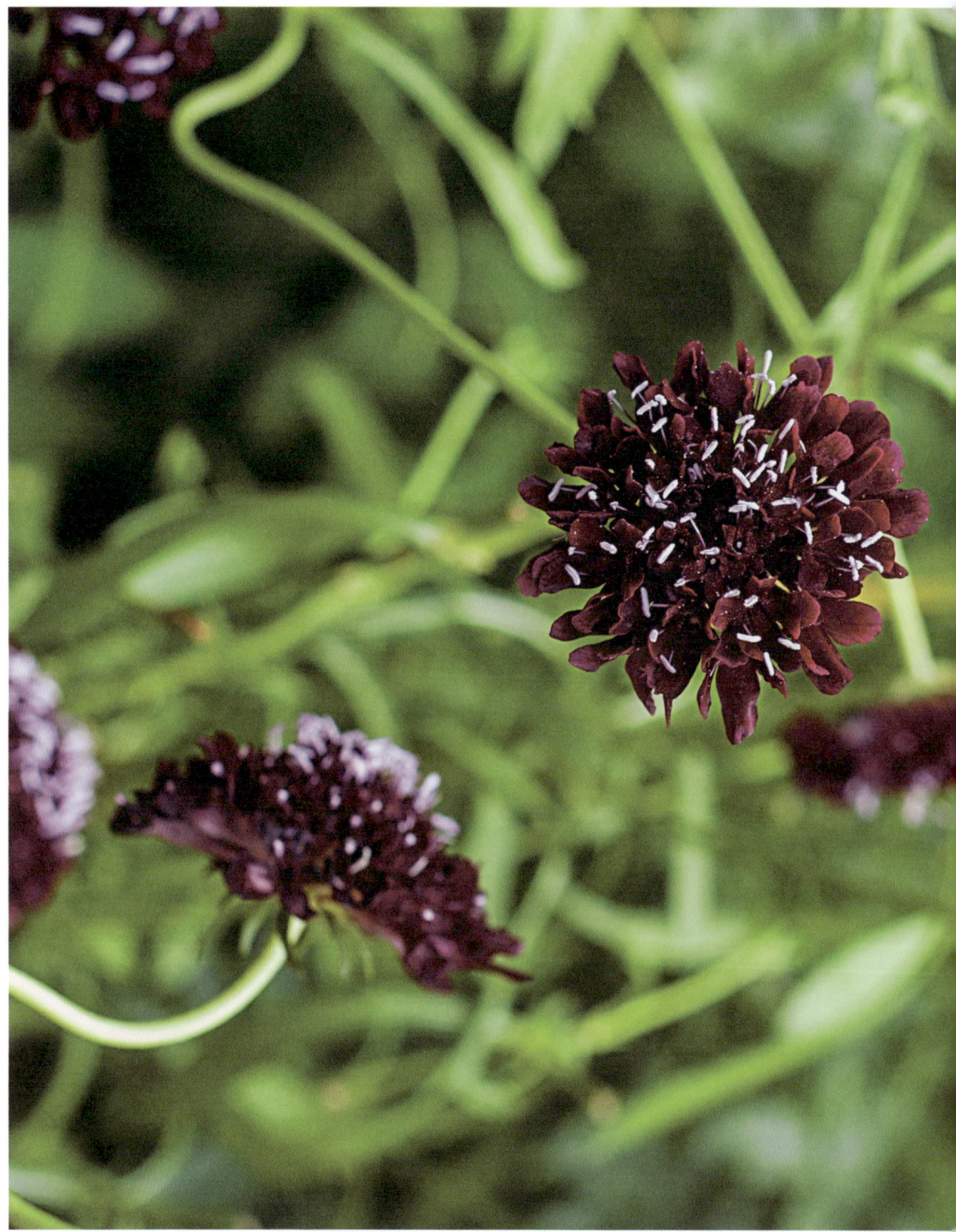

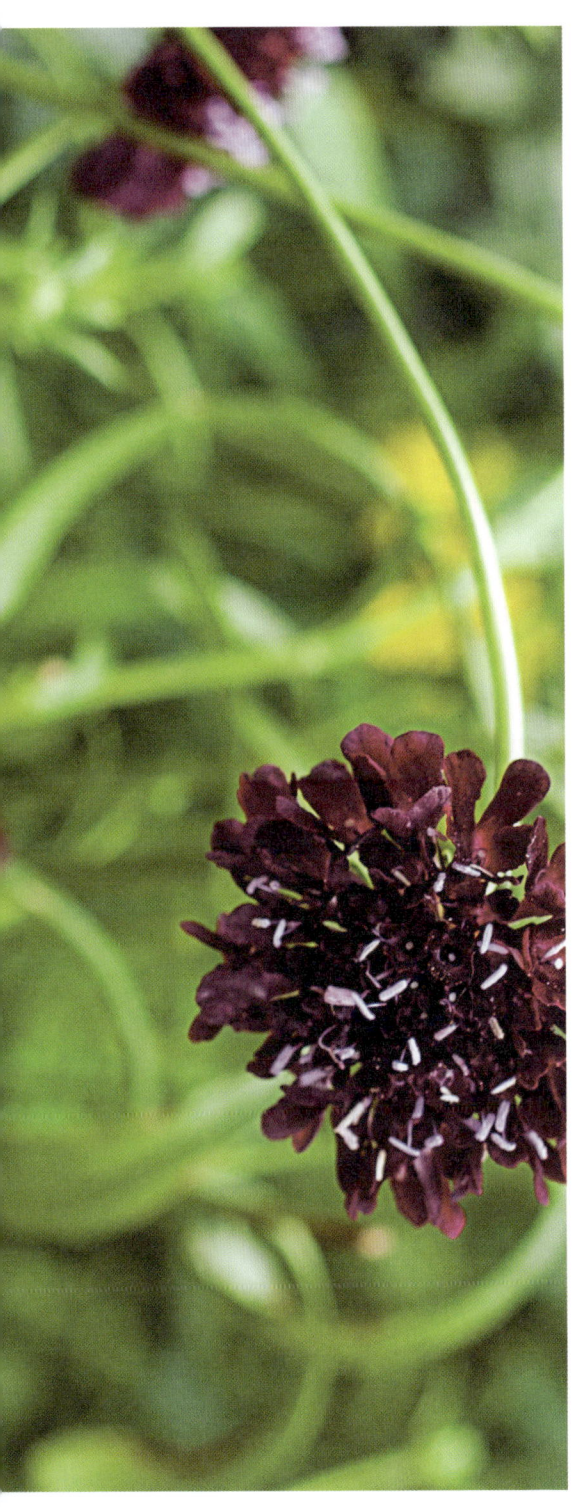

Purple Petals

While lac dye is a prized purple, many other purple dye sources grow closer to home. Common berries, garden flowers, and leaves are bursting with a biologically important class of colorants called anthocyanins. The cool hues of hibiscus, violet, iris, delphinium, hollyhock, purple cabbage, black bean, blueberries—almost all red, purple, violet, and blue plants owe their color to anthocyanins. These plants make accessible but subtle shades of muted green, soft violet, or cool gray. Be forewarned that, even with the stabilizing effect of alum, anthocyanins are fugitive colors and fade relatively quickly.

This next pigment recipe is made from the deep burgundy pincushion flower (*Scabiosa atropurpurea* "Black Knight"), a floriferous annual (shown at left). Harvest the flowers as they bloom, and if the plant isn't allowed to set seed, it will continue flowering until frost. Other blue, purple, and burgundy flowers can be used in the same way.

PINCUSHION FLOWER ICE MACERATION

While the fresh or dried blossoms of pincushion flowers can be used in a standard decoction (page 63) or tincture (page 116), here I'm sharing my method for an easy and inexpensive ice maceration. During freezing, the expansion of moisture inside the fragile petals bursts their cell walls, allowing the color to ooze out when thawed. Because no heat is applied to the dyestuff, the delicate colorants remain extra vibrant. Simply stockpile fresh blossoms in your freezer, thaw, and squeeze.

MATERIALS

· 60g fresh, richly hued pincushion flowers

· Finely woven synthetic filter cloth

· Gloves

· 10g alum

· 8g soda ash

· Distilled water, for washing pigment

METHOD

1. Place 60g fresh, plump pincushion flower heads with 30ml water in a sealed container in the freezer. Leave until frozen or you're ready to make pigment. Remove frozen pincushion flowers, enclose in a filter cloth or mesh bag, and set in a strainer over a dish to thaw.

2. Wearing gloves, squeeze and massage the flowers to release their juice, filtering it through the strainer into the dish. Compost or discard the spent dyestuff.

Recipe continued on next page

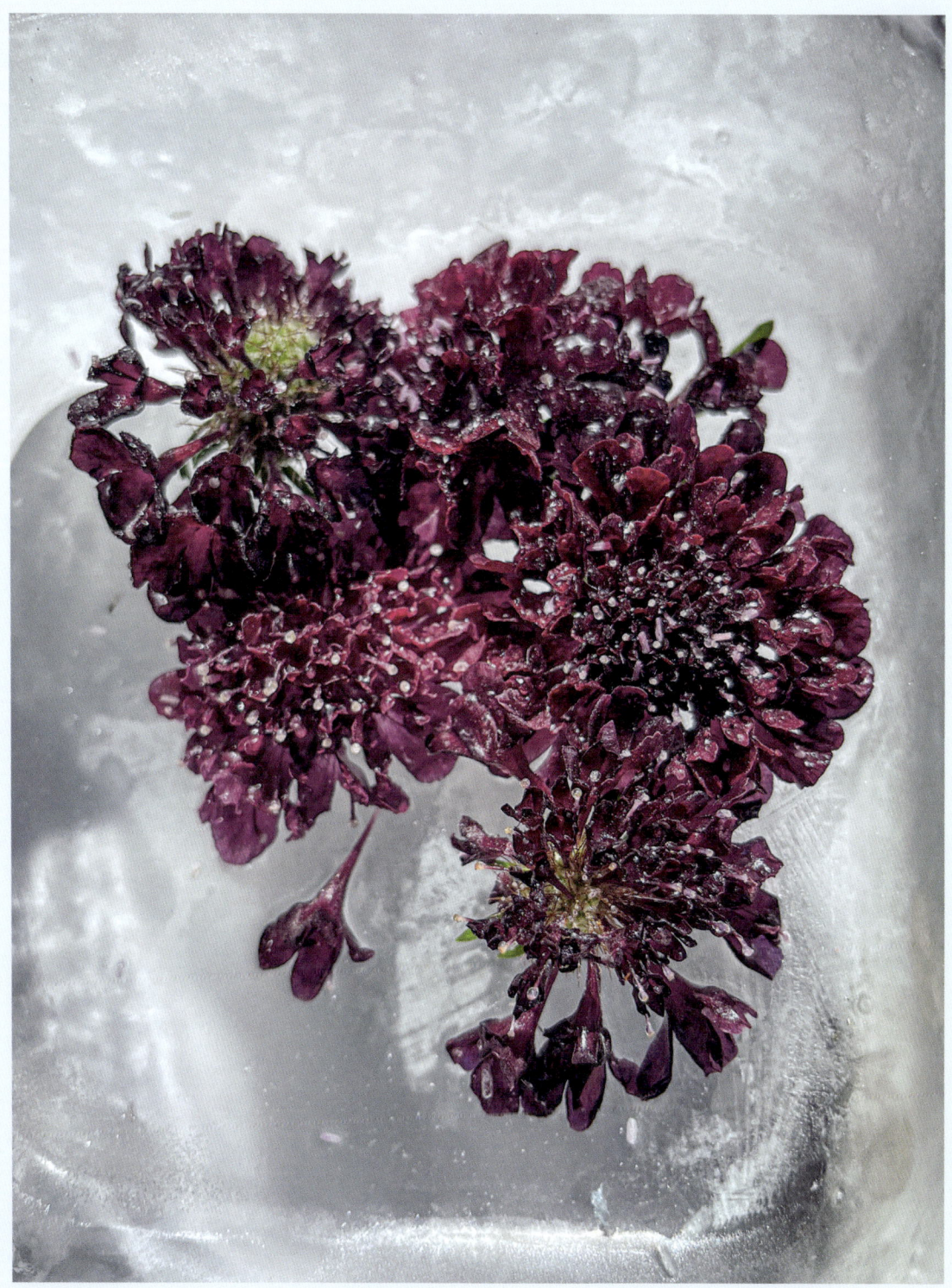

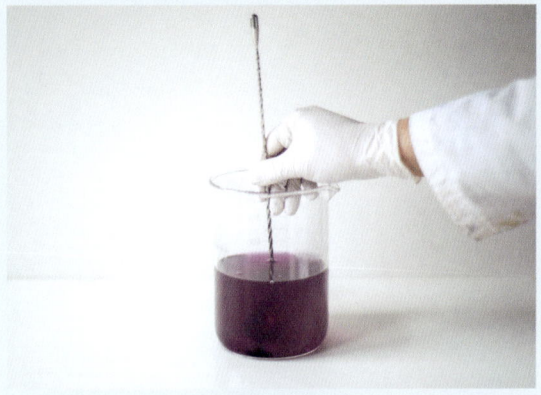

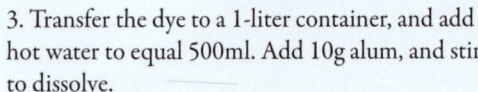

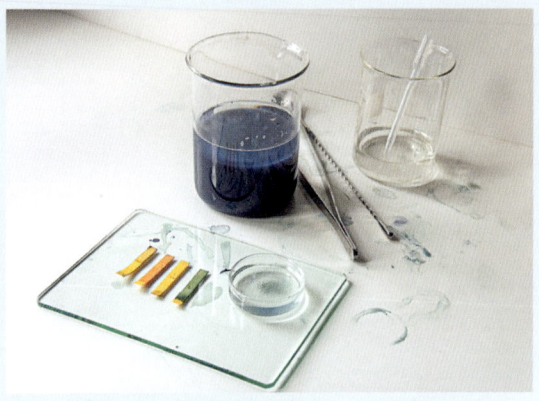

3. Transfer the dye to a 1-liter container, and add hot water to equal 500ml. Add 10g alum, and stir to dissolve.

4. Dissolve 8g soda ash in 200ml hot water. Pour 125ml of the soda ash solution into the dye mixture. Gently stir. Dip a pH test strip into the dye mixture (use tongs if necessary). Then, swish it through a small cup of distilled water. The optimum pH is 7, with wiggle room between 6 to 8. If needed, to offset this dye's innate acidity, continue adding 10ml increments of the soda ash solution, checking the pH after each addition (see page 76).

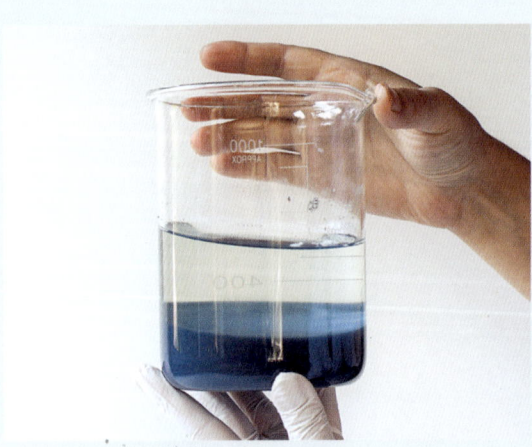

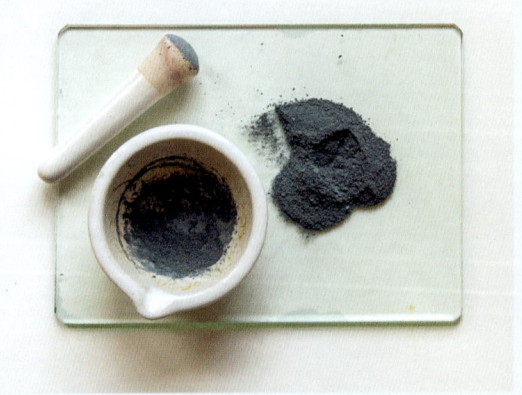

5. Filter the pigment from the supernatant. Discard the liquid, and wash the pigment following the instructions in the Master Lake Pigment recipe on page 63, starting at step 14.

ANTHOCYANINS AND PH

Anthocyanins are beguiling dyes because their hues are particularly pH sensitive, meaning the color shifts dramatically as the pH rises from magenta through purple, turquoise, and blue. In fact, a violet petal makes an excellent pH indicator strip if you happen to need one.

However, the rainbow-hued parlor trick of pH modification has limited compatability with the chemistry of lake pigments. pH-dependent lakes can only be modified across a narrow sliver of the pH scale before they destabilize and dissolve. While you may observe a flush of intriguing color as the pH falls and rises during the laking process, when the pigment forms, its color will settle in the neutral range. Though perhaps not as eye-catching as colors in the pH extremes, these shades of cool gray are beautiful in their own right.

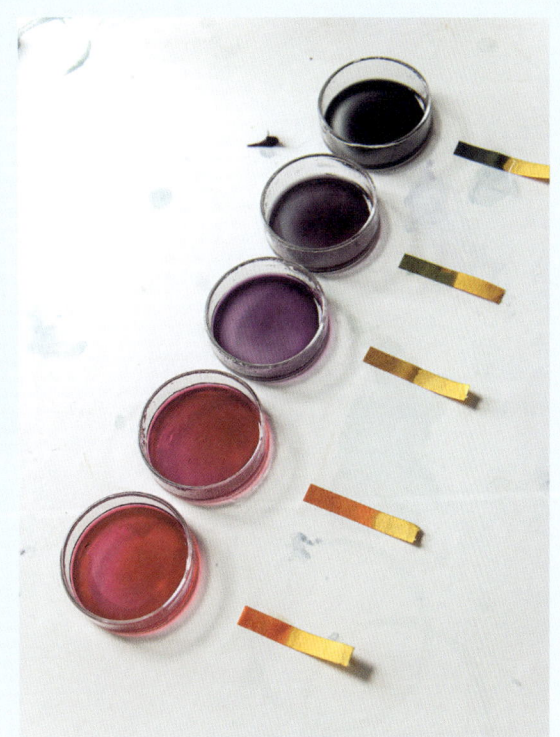

OTHER PLANTS FOR ICE MACERATION

The ice maceration method works with all juicy anthocyanin-bearing plants. Strongly hued iris, black hollyhock, and hardy hibiscus are some favorites.

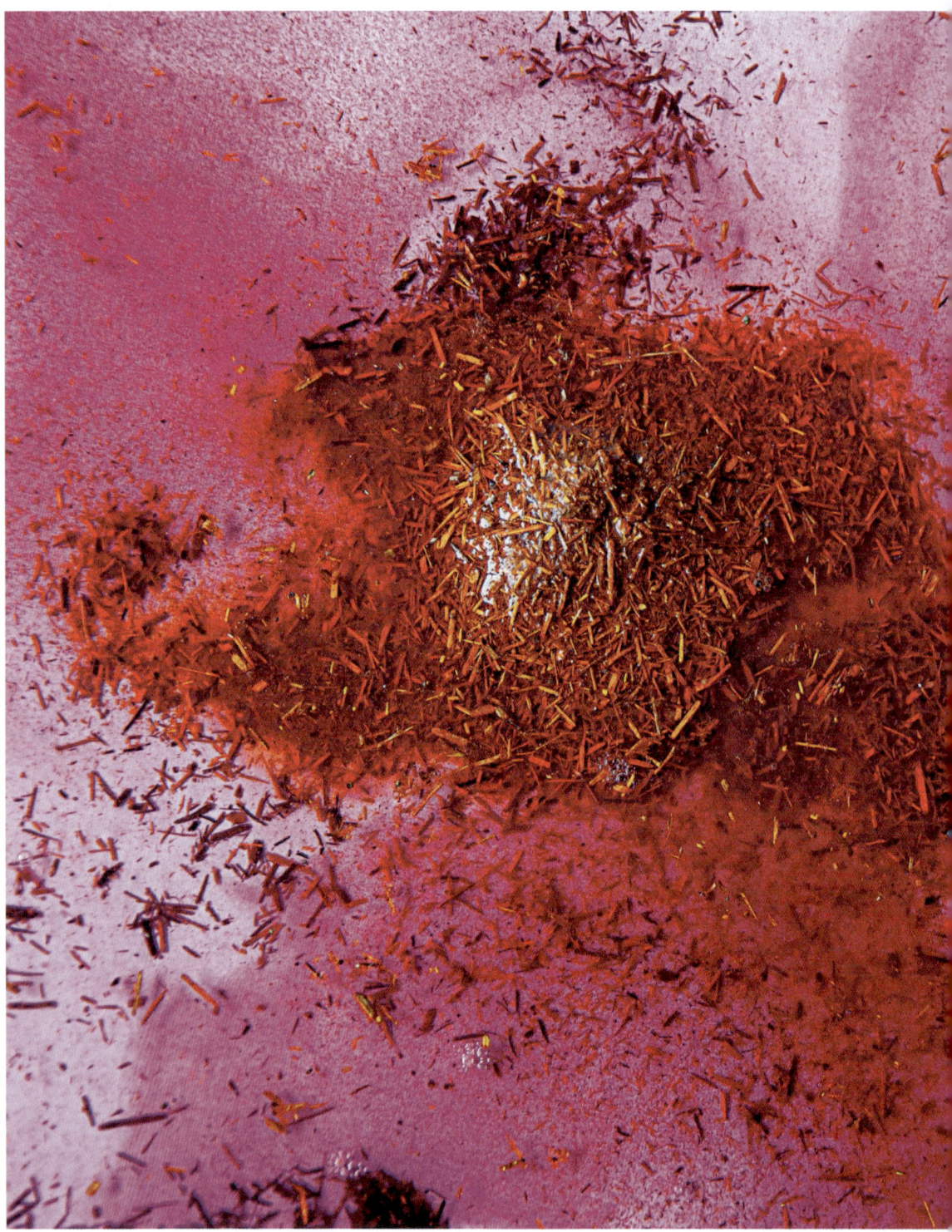

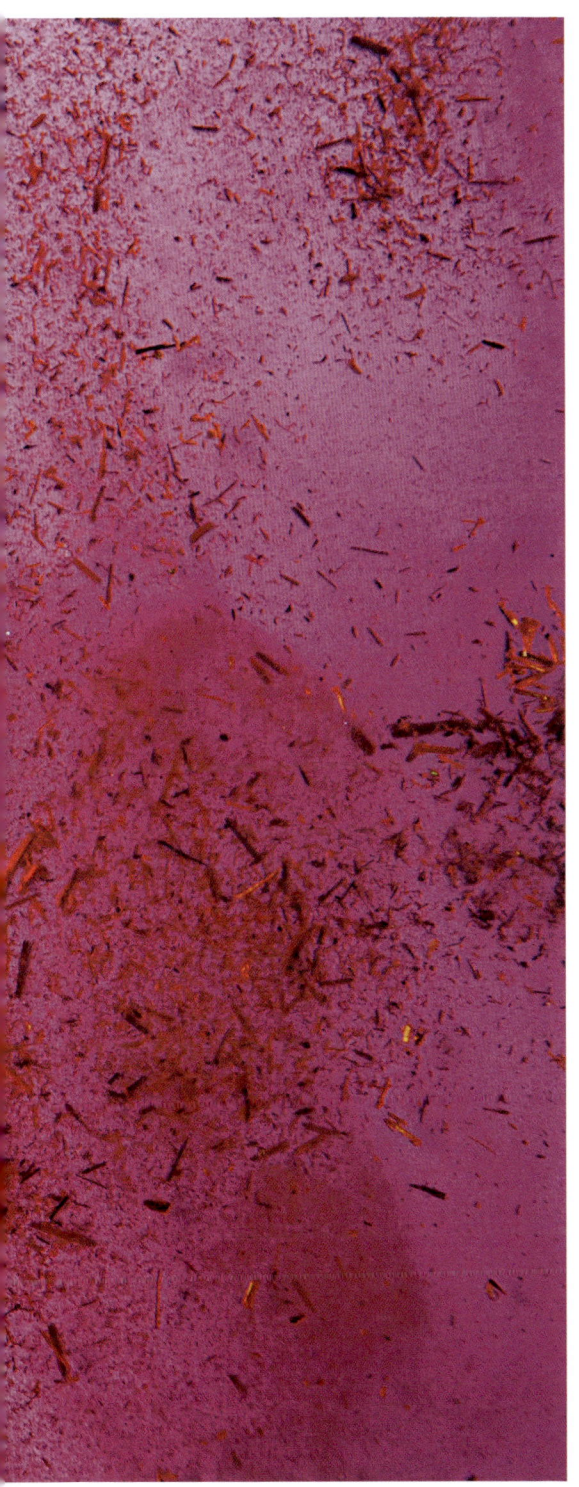

Sappanwood

For centuries, European artisans were enamored with brazilwood pigments: a host of vivid pink and purple shades from the wood of trees in the *Caesalpinia* family. Today, enthusiasm for these shades is more evident in extant pigment recipes than in artwork, because brazilwood is not terribly fast to light. However, this plant provides an example of alkaline dye extraction, a process that was often used by pigment-makers in the early modern era. This method works to strengthen and warm the shade in comparison with simple decoction. As with all botanical colors, the host of variables at play will influence the outcome.

Today, the most abundant brazilwood is the eastern variety called sappanwood (*Caesalpinia sappan*). Because commercially available sappanwood is often sold as a fine powder, it is prone to clogging the filter paper after extraction. To speed the filtering process, I recommend enclosing the powder in a very fine mesh sachet prior to extraction.

The choice of alkali is broad; soda ash solution, limewater, or wood ash water all serve the same function. Whichever you choose, handle this caustic substance with respect; keep it in a safe place, and wear gloves when handling. The key to producing a lake pigment from an alkaline extraction is balancing the pH with a commensurate quantity of acid. Don't be timid about wayfinding with pH indicator strips! If, after adding both the alum and the chalk, the pH remains above 8, you'll need a dash of something acidic. Additional alum could be used to correct the imbalance, but this would result in a paler shade of pigment. Instead, I suggest white vinegar to neutralize the mixture without creating any further precipitate, resulting in only soluble salts that will be rinsed away during washing.

SAPPANWOOD ALKALINE EXTRACTION

This order of operations, introducing the alkali before the alum, produces a somewhat gelatinous precipitate that dries in hard chunks. To ease grinding, I've included chalk in this recipe. I also recommend drying this pigment on a piece of glass or glazed ceramic, as it tends to adhere to filter paper.

MATERIALS

· 7g ground sappanwood (Eastern brazilwood)

· Mesh sachet

· 300ml alkali solution: limewater (page 43), wood ash water (page 44), or 2% soda ash solution

· 10g alum

· White vinegar

· 5g calcium carbonate

· Distilled water for washing pigment

METHOD

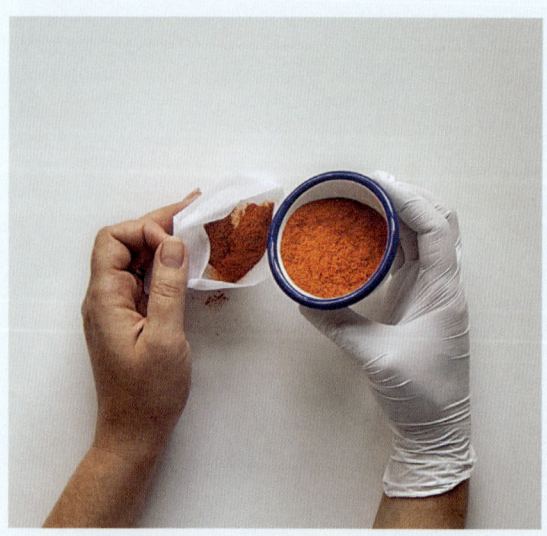

1. Enclose 7g dry, finely ground or shaved sappanwood in a sachet.

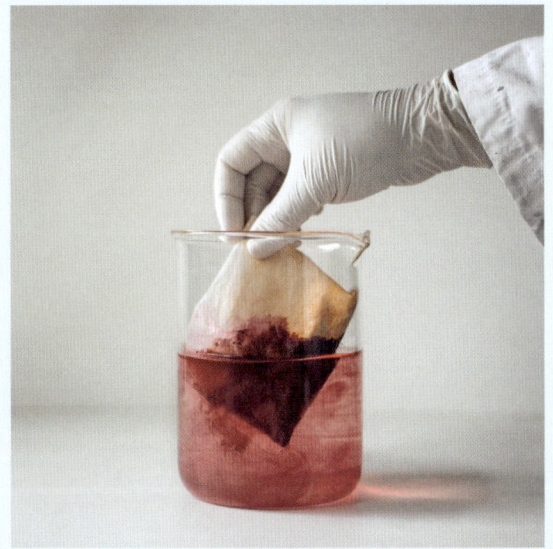

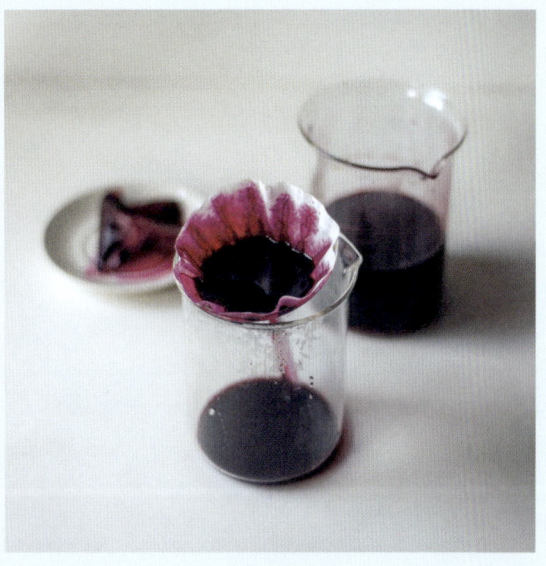

2. Place the sachet in 300ml alkaline solution. Steep sappanwood for at least six hours at room temperature, or up to one week. Optional: Bring to a simmer for 30 to 60 minutes, topping up the water as needed to keep dyestuff submerged. This extracts more color, but it can also desaturate its hue.

3. Wearing gloves, remove the sachet and compost or discard its contents. Filter the dye to remove every speck of plant material.

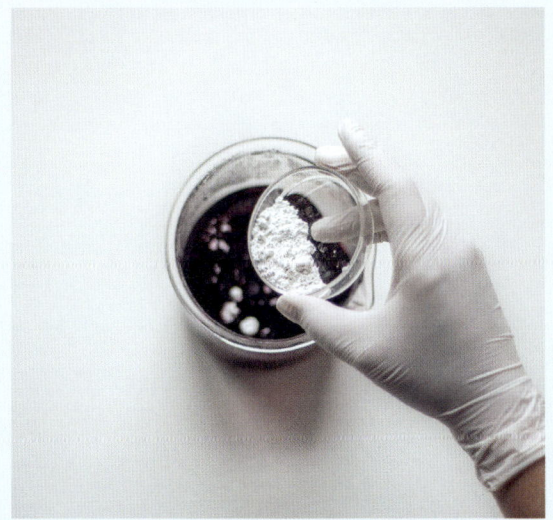

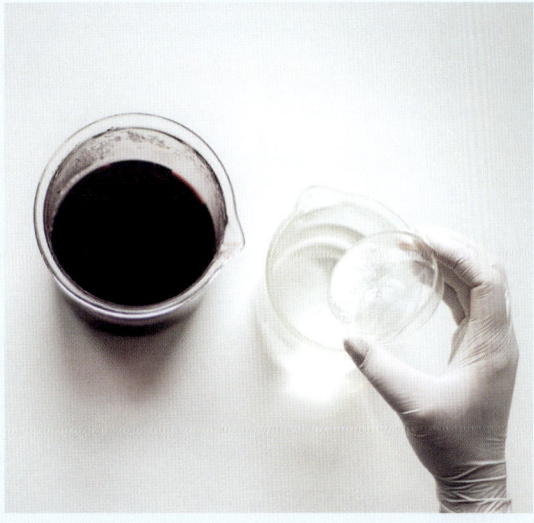

4. Transfer the dye to a 1-liter container, and add hot water to equal 500ml if needed. Stir in 5g calcium carbonate.

5. In a separate container, dissolve 10g alum in about 200ml hot water.

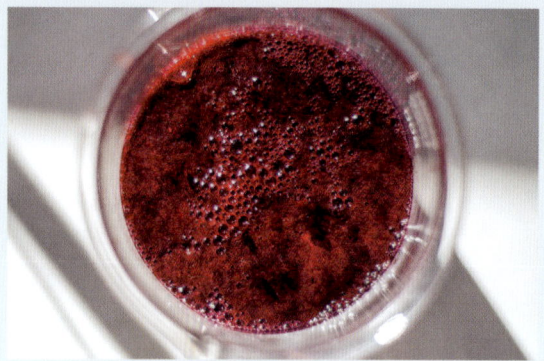

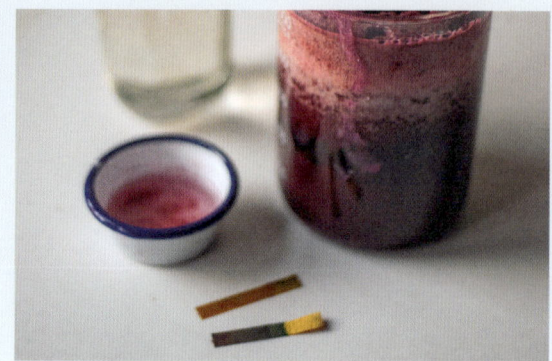

6. This is an extra-ebullient neutralization reaction, so take it slow. Pour about one-quarter of the alum solution into the dye mixture. Gently stir. Repeat with the remaining three-quarters, one at a time.

7. The pH can vary depending on your dye source, alkali strength, and steep time. Check the pH by dipping a pH test strip into the dye mixture (use tongs if necessary). Then, swish it through a small cup of distilled water to remove any pigment obscuring the color. Compare the color to the manufacturer's key. The optimum pH is 7, with leeway between 6 to 8. If the pH is above 8, add a little dash of white vinegar to the lake mixture, stir, and test again. Repeat as needed until the pH is neutral.

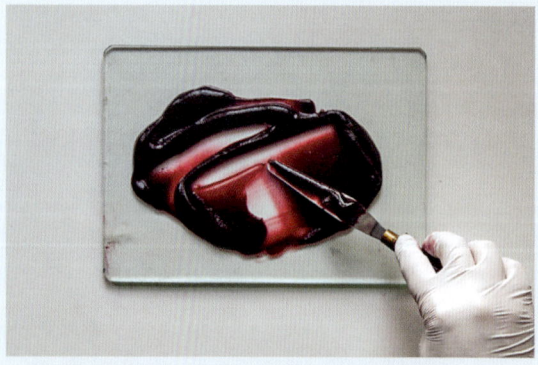

8. Filter the pigment from the supernatant. Discard the liquid and wash the pigment following the instructions in the Master Lake Pigment recipe on page 63, starting at step 14. If drying to a powder, spread the wet pigment paste on a piece of glass or glazed ceramic.

OTHER PLANTS FOR ALKALINE EXTRACTION

Woody and tannin-rich dye sources like birch bark (*Betula* spp.), buckthorn bark (*Rhamnus* spp.), stone fruit tree barks (*Prunus* spp.), and logwood (*Haematoxylum campechianum*).

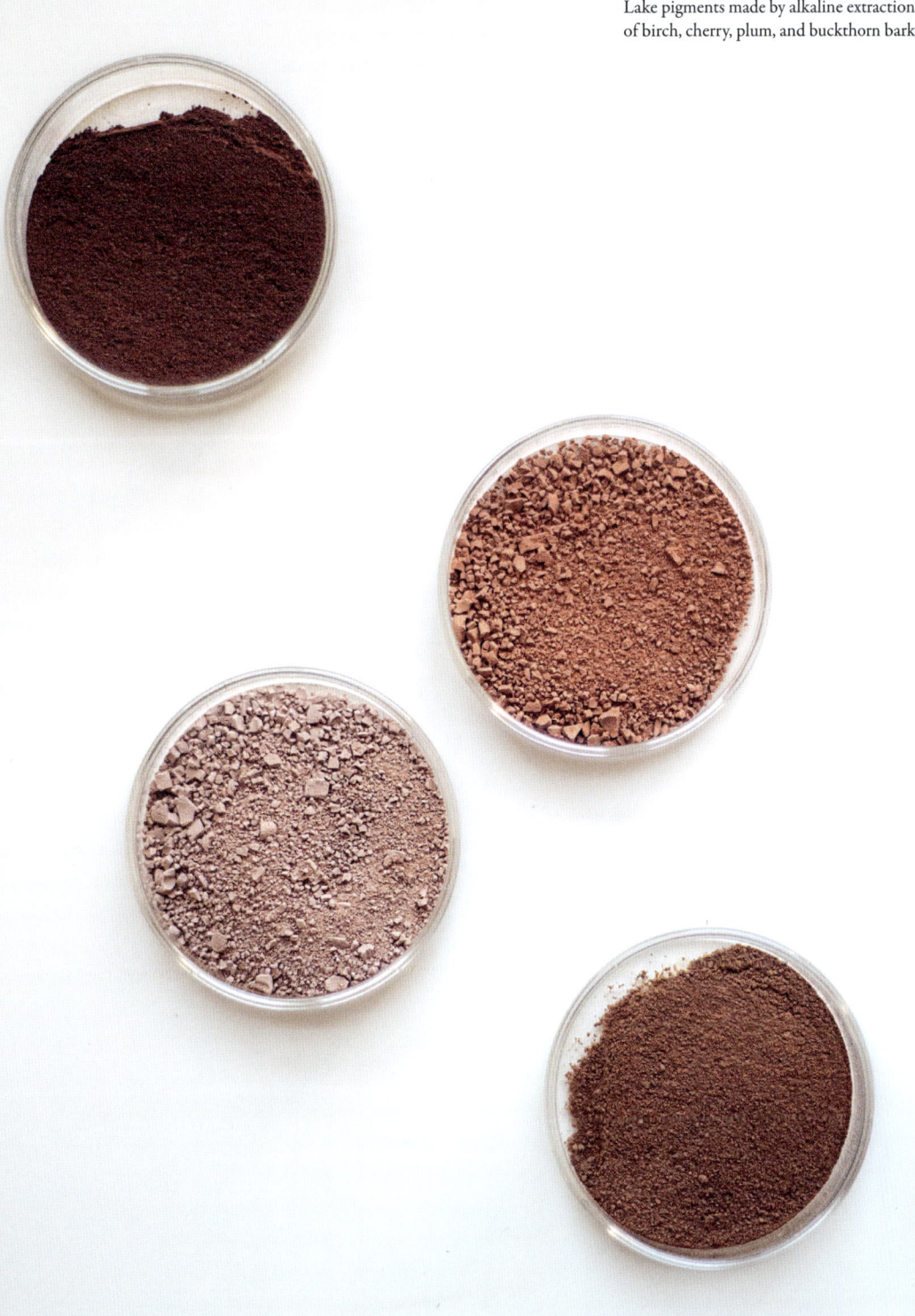

Lake pigments made by alkaline extraction
of birch, cherry, plum, and buckthorn barks.

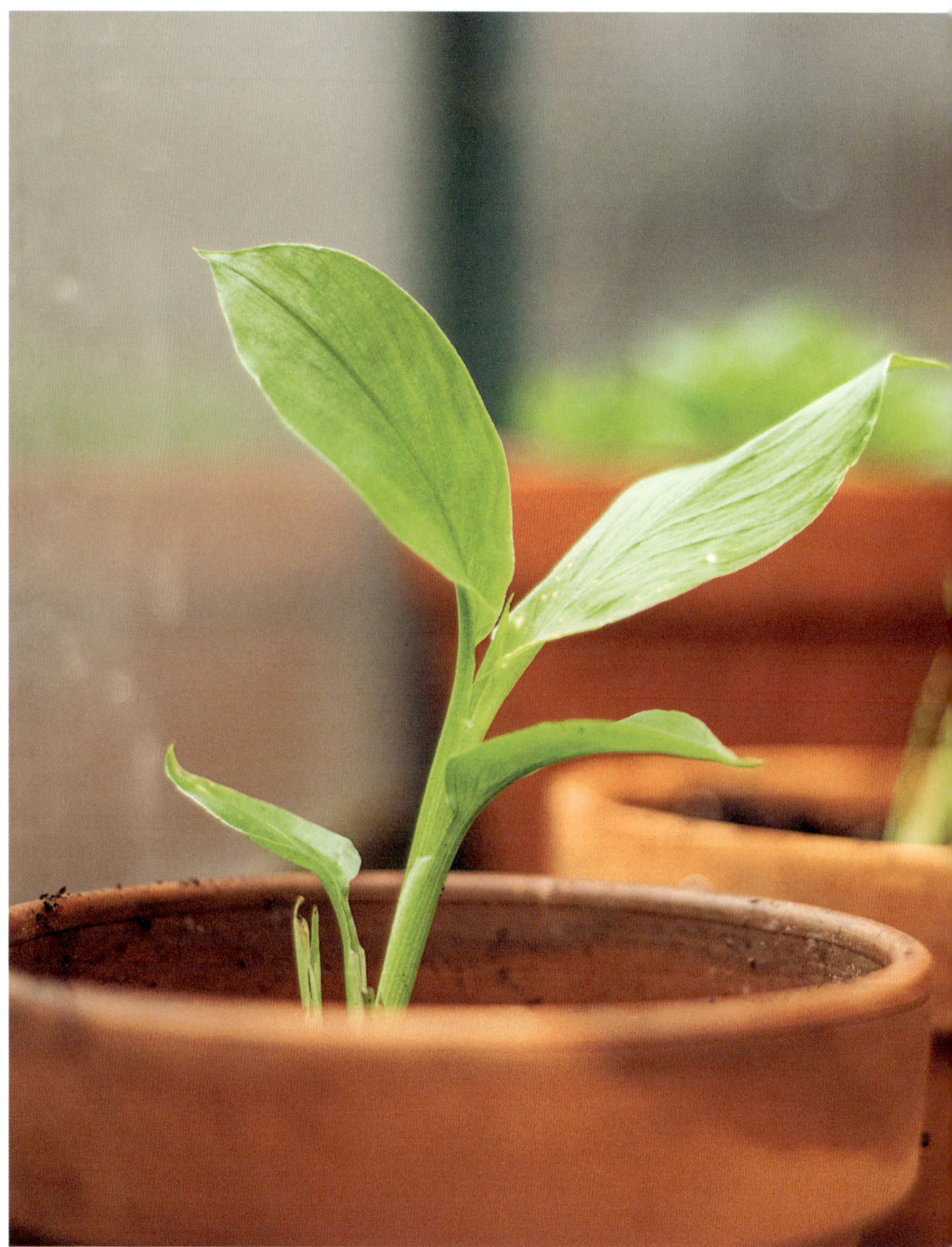

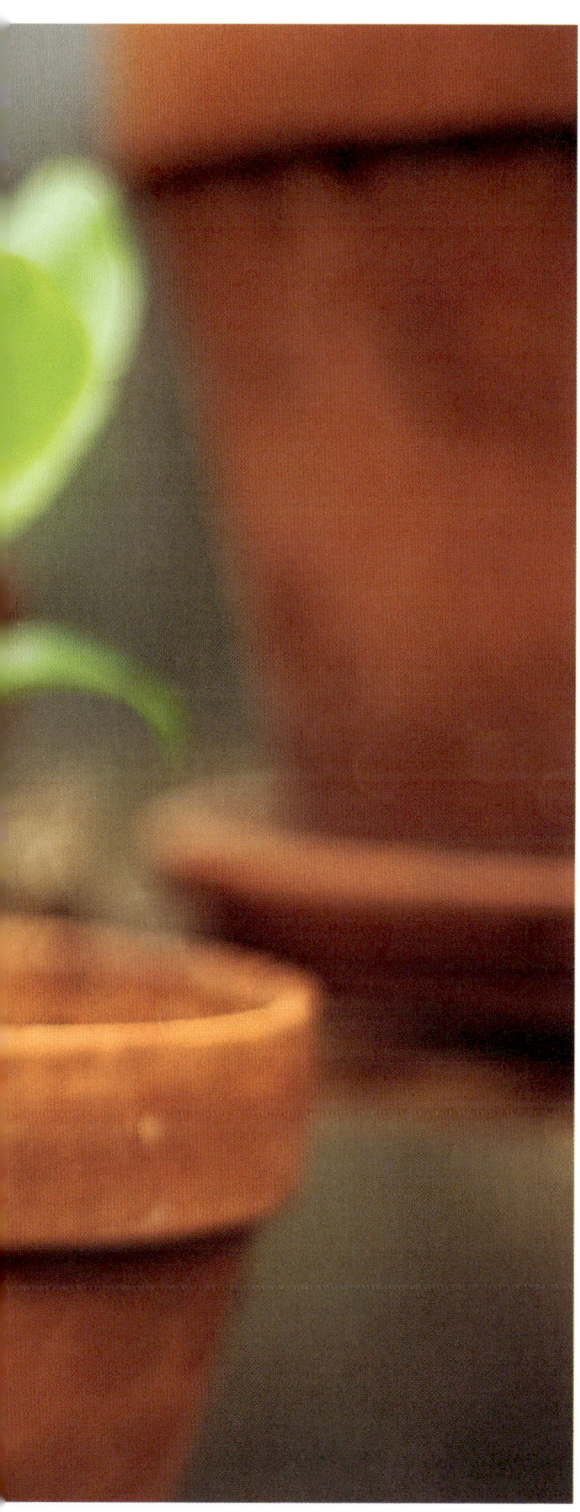

Turmeric

Turmeric (*Curcuma longa*) is a tropical perennial grown from rhizomes which contain the flavorful, healthful, and colorful compound curcumin. This colorant is soluble in alcohol, as demonstrated in the following recipe. You can source turmeric powder from a dye vendor or grocery store spice aisle, or you may be able to find fresh rhizomes stocked in the vegetable department. To grow your own turmeric in less-than-tropical conditions, seek out rhizomes from a nursery who can ensure that it's disease-free and ready to grow, along with guidelines for success in your local conditions. These rhizomes are generally available for pre-order in winter and arrive in springtime. Given ample heat and light indoors or in a greenhouse, your turmeric will soon sprout shoots. Grow it under cover, or outdoors once the soil temperature tops 16°C (60°F). Provide turmeric with plenty of nutrition and water, and wait until the last moment before frost to harvest the largest possible crop of rhizomes. Save a few healthy rhizomes over winter to grow on and perpetuate your stock the following spring, storing them just as you would dahlia tubers.

Similar to turmeric, the heartwood of several trees is best extracted by alcohol. These resinous, water-insoluble dyes come from sandalwood and barwood, among other redwoods. (See Other Plants for Extraction by Alcohol on page 117). Each of these redwoods has a history and a heritage. Like brazilwood, sandalwood has historically been overharvested to the point of endangerment. Today, preservation efforts in its native India allow for the certified export of limited amounts. If you choose to purchase any of these woods, source your dye material from a reputable specialist (page 171). Seek out the Safety Data Sheet from your vendor to learn about the species of redwood you are working with and where it comes from.

TURMERIC TINCTURE

Some biochromes dissolve more readily in alcohol than water. One such source is turmeric, which can be used in a standard decoction but provides significantly more color via alcohol tincture.

MATERIALS

· 5g ground turmeric rhizome, dry
(or 25g fresh, minced)

· 100ml grain alcohol or vodka (preferably 140 to 190 proof)

· 10g alum

· 5g soda ash

· Distilled water for washing pigment

METHOD

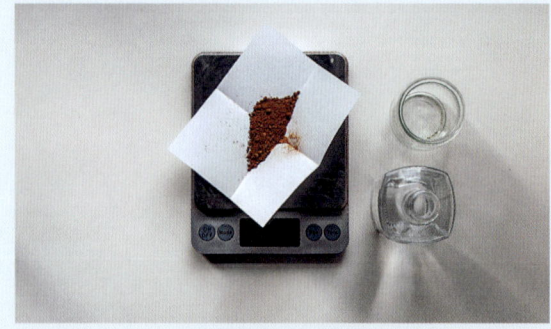

1. Measure 5g ground turmeric into a small jar with a tightly fitting nonmetallic lid.

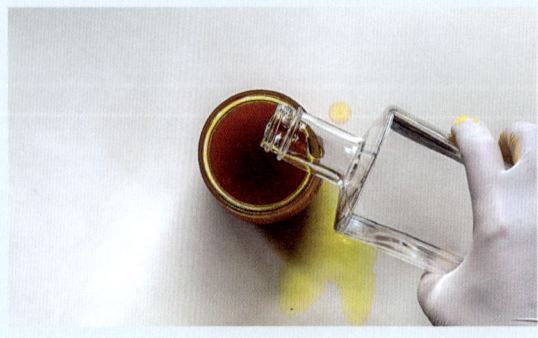

2. Cover the turmeric with 100ml grain alcohol. Cap the jar, and store in a safe spot out of direct sunlight.

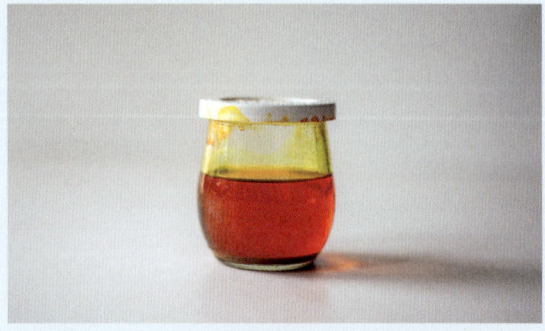

3. Macerate for one to two weeks, shaking the jar once every two to three days. Longer maceration doesn't hurt!

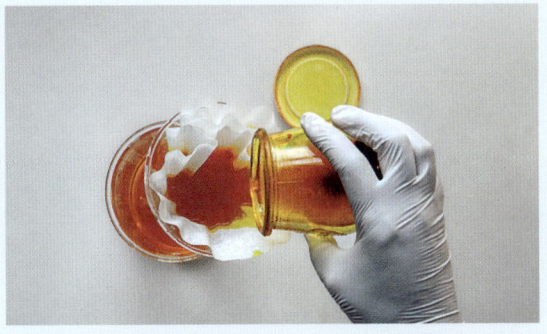 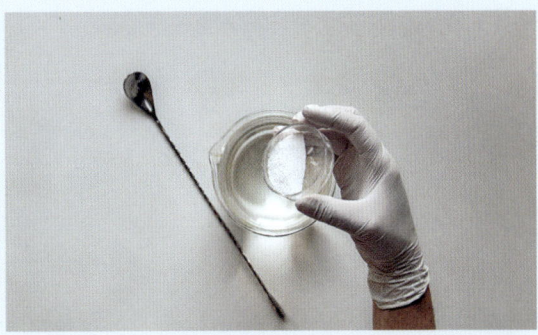

4. Filter the mixture; the liquid that passes through the filter is the tincture. Compost or discard the turmeric powder retained by the filter paper. Transfer the tincture to a 1-liter container.

5. In a separate container, add 10g alum to 400ml hot water. Stir until dissolved. Pour the alum mixture into the turmeric liquid. You might notice the liquid turn milky.

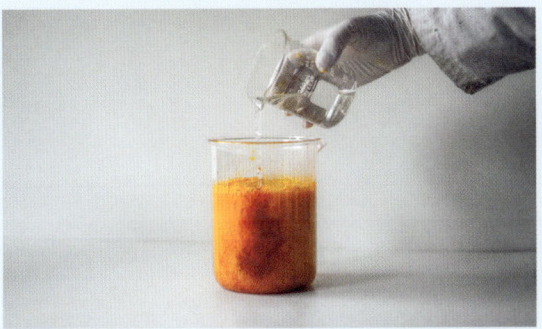 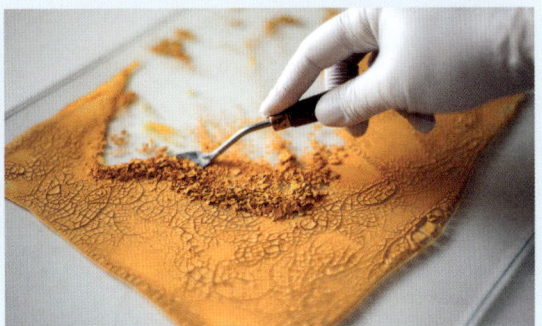

6. In a separate container, dissolve 5g soda ash in about 200ml hot water. Pour roughly half the soda ash solution into the dye mixture. Gently stir. Repeat with the remaining soda ash. You might see the turmeric initially flush red from the alkalinity, and then stabilize.

7. Filter the pigment from the supernatant. Discard the liquid, and wash the pigment following the instructions in the Master Lake Pigment recipe on page 63, starting at step 14. Use an alcohol swab to easily clean residue from your tools.

OTHER PLANTS FOR EXTRACTION BY ALCOHOL

Sandalwood (*Pterocarpus santalinus*), narrawood (*Pterocarpus indicus*), barwood (or African padauk) (*Pterocarpus soyauxii*), and camwood (*Baphia nitida*) can all be treated with an alcohol extraction. Anthocyanin-bearing plants are also suitable.

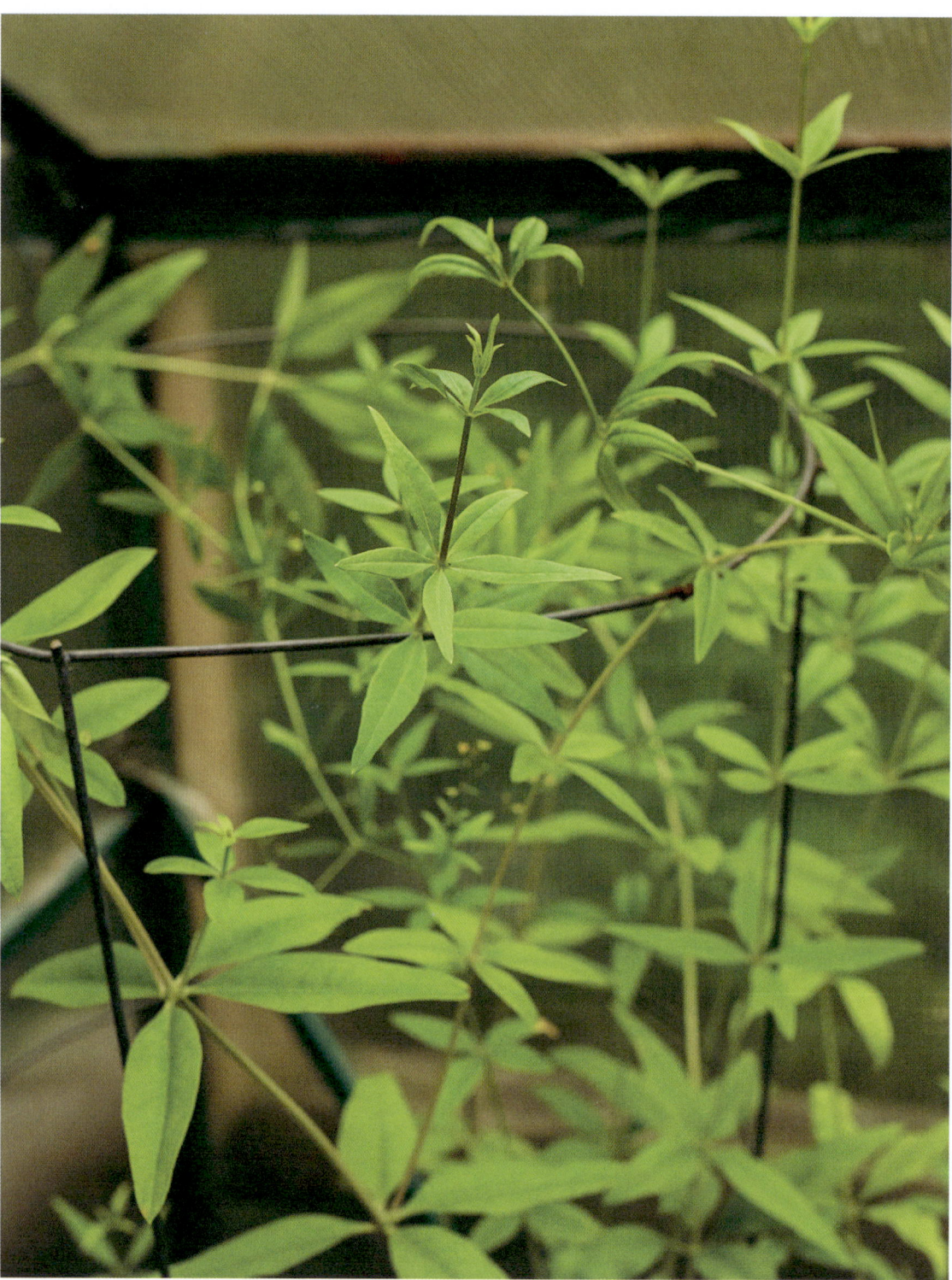

Madder Root

The history of madder dye stretches deep into the past, and it is seen in varying shades of crimson, pink, brick red, orange, and purple. The roots of madder (*Rubia tinctorum*) contain at least 15 different dye colorants including alizarin, making it a bounteous but sometimes bewildering dye. (The commercially available alizarin crimson you might already have in your paint box is the synthetic copy of madder's naturally occurring red molecule.) While the standard hot water extraction provides a range of lovely hues, it does not access the full potential of this plant. Tinkering with each variable of the process—water source, pH, temperature, and duration—reveals the nuance of this plant. I have found that an acidic fermentation, followed by a moderately heated extraction, has twice the tinctorial strength of a hot water extraction alone and provides a range of glowing reds.

In the garden, madder is an easygoing herbaceous perennial with an almost weedy habit. It sprawls across the ground and scrambles up neighbors much like its cousins the woodruffs and bedstraws, which also contain dyes (albeit in far less concentration). Madder enjoys a sunny position on fertile but sharply drained soil, and it benefits from stakes or trellising to train it upward rather than outward. (Beware that the leaves have minute hooked spines to aid its clambering, so always wear gloves around this plant.) Madder's color potential is improved by sweetening acidic soil with fireplace ashes or lime. It takes two to five growing seasons for the roots to reach the size for harvesting, at least the circumference of a pencil. In areas that see winter temperatures below -12°C (10°F), a layer of protective mulch or overwintering under cover is advisable. (After twice losing my madder plants to cold and damp, I now grow it trained up a tomato cage in an unheated greenhouse.)

After digging the roots, rinse off dirt and remove the brown bark. Chop the roots into short segments and dry them. During this stage, enzymes within the roots release colorants from their precursor molecules. This process of cleaving dye from glucose is known as hydrolysis and can be encouraged during processing and extraction. Historically, dye yield was increased by steaming or oven-drying the roots, allowing maximum color to develop. It might be worth experimenting with these methods, or you can simply spread the roots out on screens to dry. Those same sugars that make madder an easy plant to ferment also make it easy prey for mold, so do this in a warm, well-ventilated area. Of course, while you're waiting a few years to grow and process your own madder harvest, you can source dried whole or pulverized madder roots from dye vendors.

WORKING WITH WHOLE OR GROUND MADDER ROOT

Whole root segments can be left to swim free; however, if you're using pulverized roots, enclose them in a fabric sachet for fermentation, leaving plenty of room for the roots to swell as they hydrate. The fabric bundle will keep them submerged (not floating = not moldering), but it might also prevent the bubbles produced by the fermentation from escaping. Each day when you check on your ferment, gently push down on the fermentation weight to express the air bubbles trapped in the fabric.

MADDER ROOT FERMENTATION EXTRACTION

This fermentation extraction enlists a legion of microbes to gradually break down madder's woody structure and digest its sugars, releasing color through hydrolysis. With this recipe, fermentation is ensured by culturing the liquid with a dash of sauerkraut juice or other living lacto-ferment, seeding the roots with microbes to get a head start over mold. With patience, you'll soon observe bubbles rising through the roots. Municipal water purposefully inhibits microbes, so make sure to use well water, gathered rainwater, or distilled water to support a healthy fermentation.

MATERIALS

· 30g dried madder root (*Rubia tinctorum*), whole or ground (see Working with Whole or Ground Madder Root, p. 119)

· ½ tablespoon raw sauerkraut juice or other unpasteurized lacto-ferment liquid

· Well water, rainwater, or distilled water for fermentation

· Two grades of filter:
   Fine: synthetic mesh, such as dashi sachets or nut milk strainers
   Superfine: 20-cm (8-in) round paper coffee filters

· 10g alum

· 10g soda ash

· Distilled water, for washing pigment

METHOD

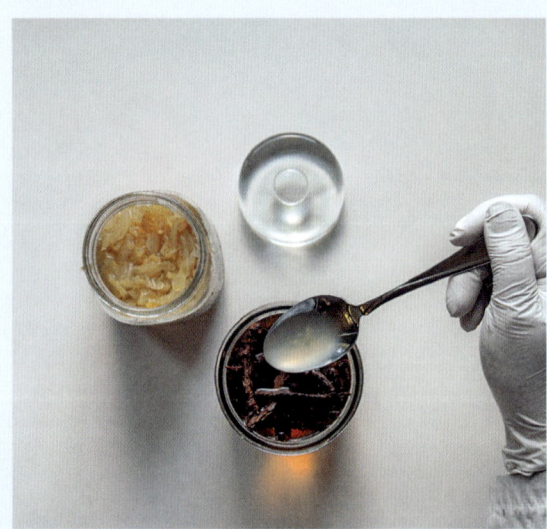

1. Weigh out 30g dried madder roots in a jar or beaker. Douse the roots with roughly 1L boiling water. Steep for five minutes to neutralize mold spores and draw out off-colors. Filter through a fine mesh, and discard the brownish liquid. Return the roots to a clean jar measuring 0.5 to 1 liter (1 pint to 1 quart). Cover madder roots by one to two inches with roughly 300 to 400ml room temperature, fermentation-supporting water. Add ½ tablespoon sauerkraut juice or lacto-ferment liquid of your choice.

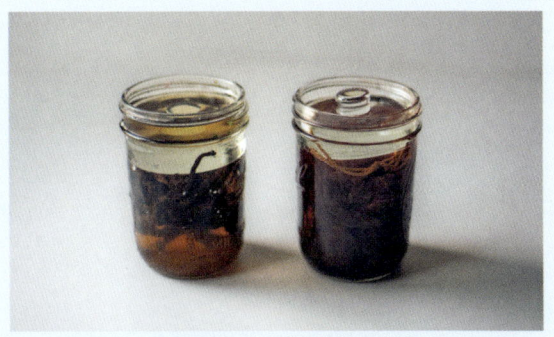

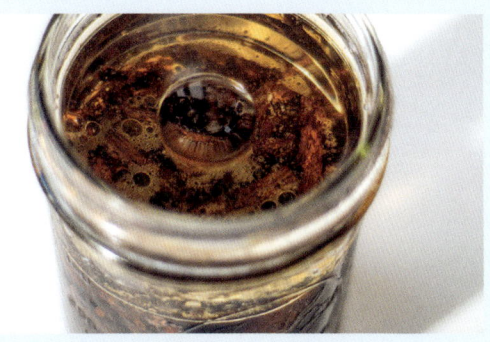

2. As with fermenting vegetables, sinking the plant material below the water's surface immerses it in the anaerobic conditions that mold cannot reach. Weigh down the madder roots with a glass fermentation weight or similar. It might take a little creative problem-solving to find an object (such as an inverted jar lid filled with water) that covers as much surface area as possible and still allows the carbon dioxide created by the fermentation to escape. You can also stretch a scrap of fabric or a coffee filter over the mouth of the jar to keep out dust and critters.

3. Leave the madder at room temperature, about 18 to 21°C (65 to 70°F), in a shady spot out of direct sunlight. After a day or two, you should see bubbles amidst the roots and ringing the surface.

Check the madder daily, adding water if the level drops. Allow it to ferment for one to two weeks. During this time the pH should be about 5, and the scent should be a combination of damp earth and tangy-sweet vinegar. If the fermentation looks and smells healthy, you can keep it going even longer until the bubbles cease and the pH rises toward neutral, indicating that all the sugars have been consumed. If mold begins forming on the surface, scrape it off and stay the course. If mold becomes significant and persistent, move to the next step.

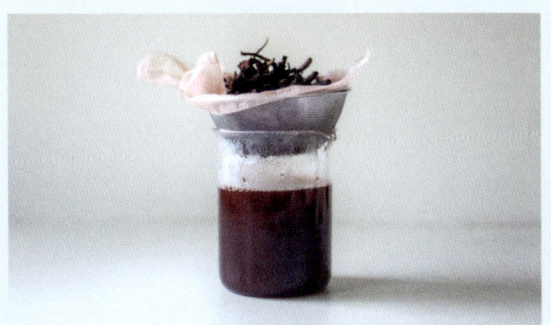

4. The madder has now been predigested by the fermentation, but it still requires hot water extraction. The fermentation liquid tends to contain brownish-pink colorants, so I prefer to separate this liquid and make the hot extraction in fresh water. Filter the madder roots from the fermentation liquid using a fine synthetic mesh, preserving all the bits and dregs of dyestuff. Optional: The fermentation liquid can be heated and laked itself using 5g alum and 3g soda ash. Otherwise, compost or discard this liquid.

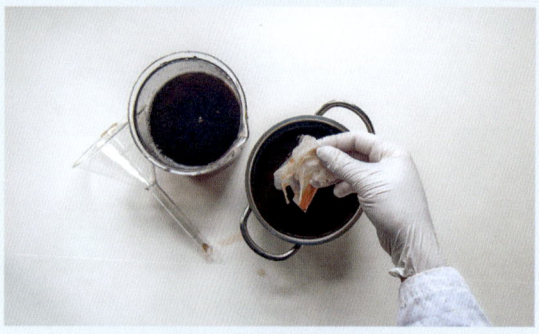

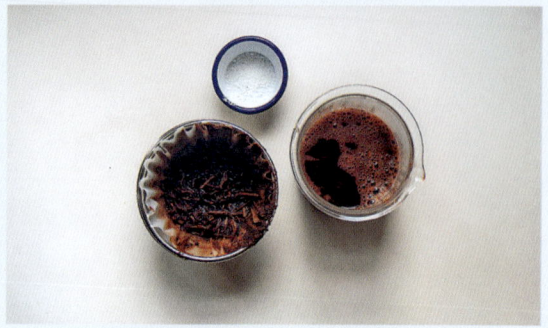

5. Transfer the fermented madder roots to a 1-liter dye pot of fresh water. Gather up the filter and swish it through the water too, releasing every bit of madder sediment.

6. Heat the madder to 66 to 76°C (150 to 170°F) for 30 to 60 minutes. This is steaming hot but below a simmer. Herbalists and alchemists term this low, slow process "a digestion." A laboratory hotplate with a magnetic stirrer can be a worthwhile investment here.

7. It is important to filter this extraction while it's hot and the alizarin is in solution. Filter the hot dye through a fine synthetic mesh to remove every scrap of madder roots. Compost the roots, or save them for a secondary extraction if you doubt that they are exhausted.

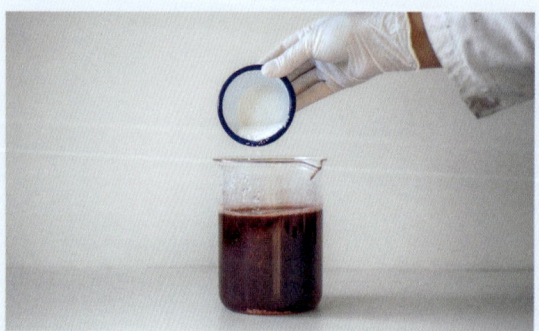

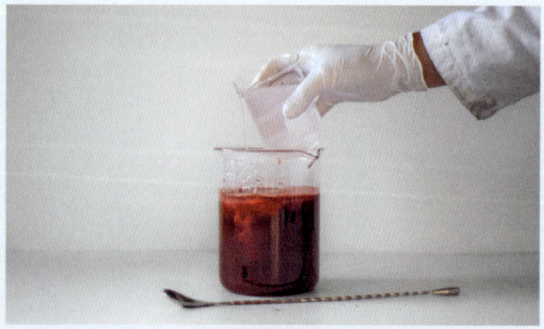

8. Add water to equal 500ml, if needed, and heat the dye to 43 to 76°C (110 to 170°F). Add 10g alum to the hot dye, stirring until dissolved. You might notice the colorant coalescing.

9. Dissolve 10g soda ash in 200ml hot water. Pour 100ml (half) of the soda ash solution into the madder dye.

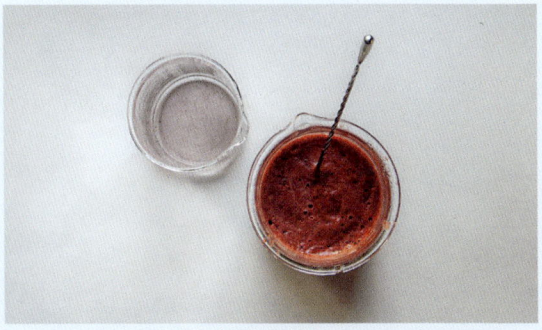

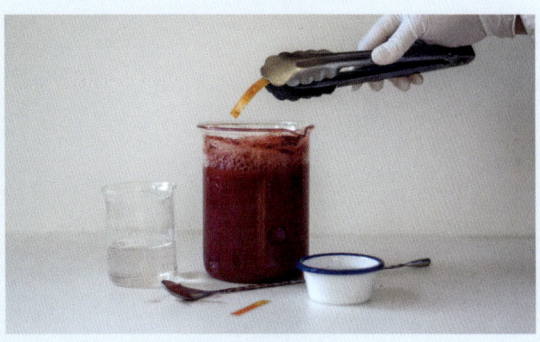

10. Because the fermentation has created a low pH, additional soda ash might be needed to neutralize and precipitate the lake. Add 20ml of the remaining soda ash solution, and gently stir.

11. Test the pH and repeat as needed until the pH reaches 7, with leeway of 6 to 8.

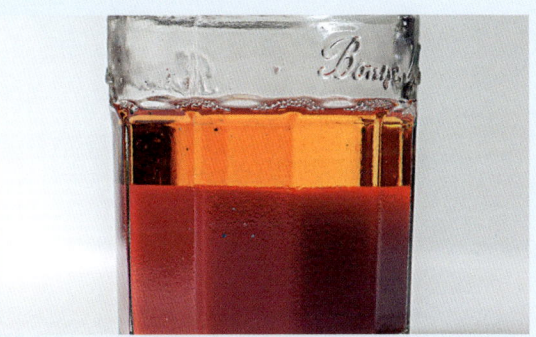

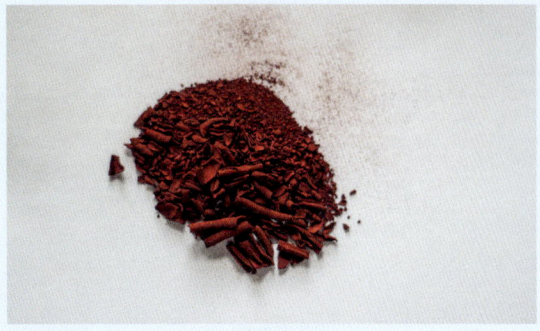

12. Switch now to superfine paper coffee filters, and filter the pigment from the supernatant. Note: It is normal for the supernatant to retain an amber shade from non-mordant colorants.

13. Discard the liquid, and wash the pigment following the instructions in the Master Lake Pigment recipe on page 63, starting at step 14.

OTHER PLANTS FOR EXTRACTION BY FERMENTATION

Rubia cordifolia, *R. akane*, and *R. peregrina* are other species of madder. The roots of bedstraws, including *Galium verum*, *G. aparine*, *G. saxatile*, and *G. odoratum*, might be worth exploring with this technique too.

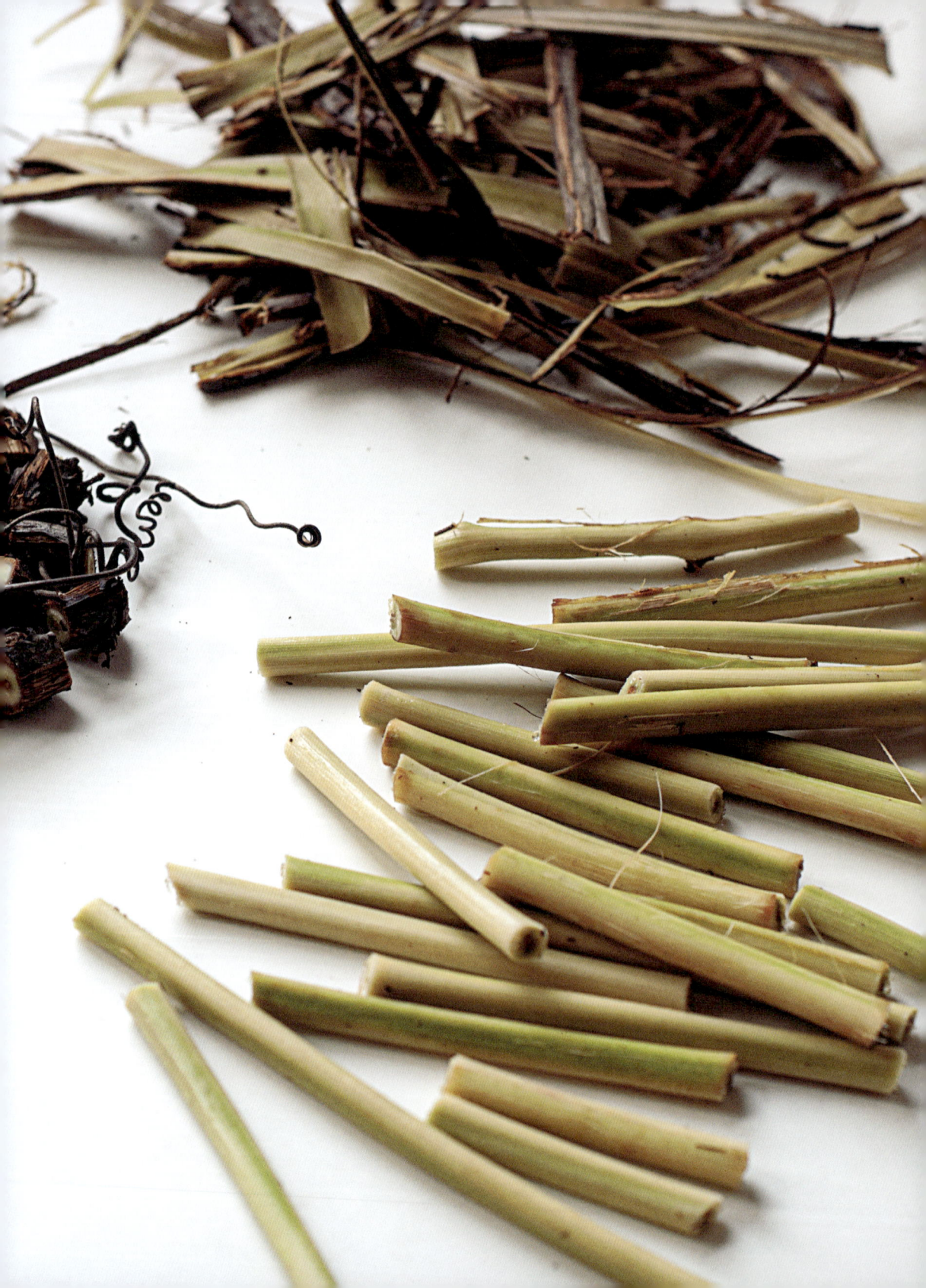

4

OTHER PLANT PIGMENTS

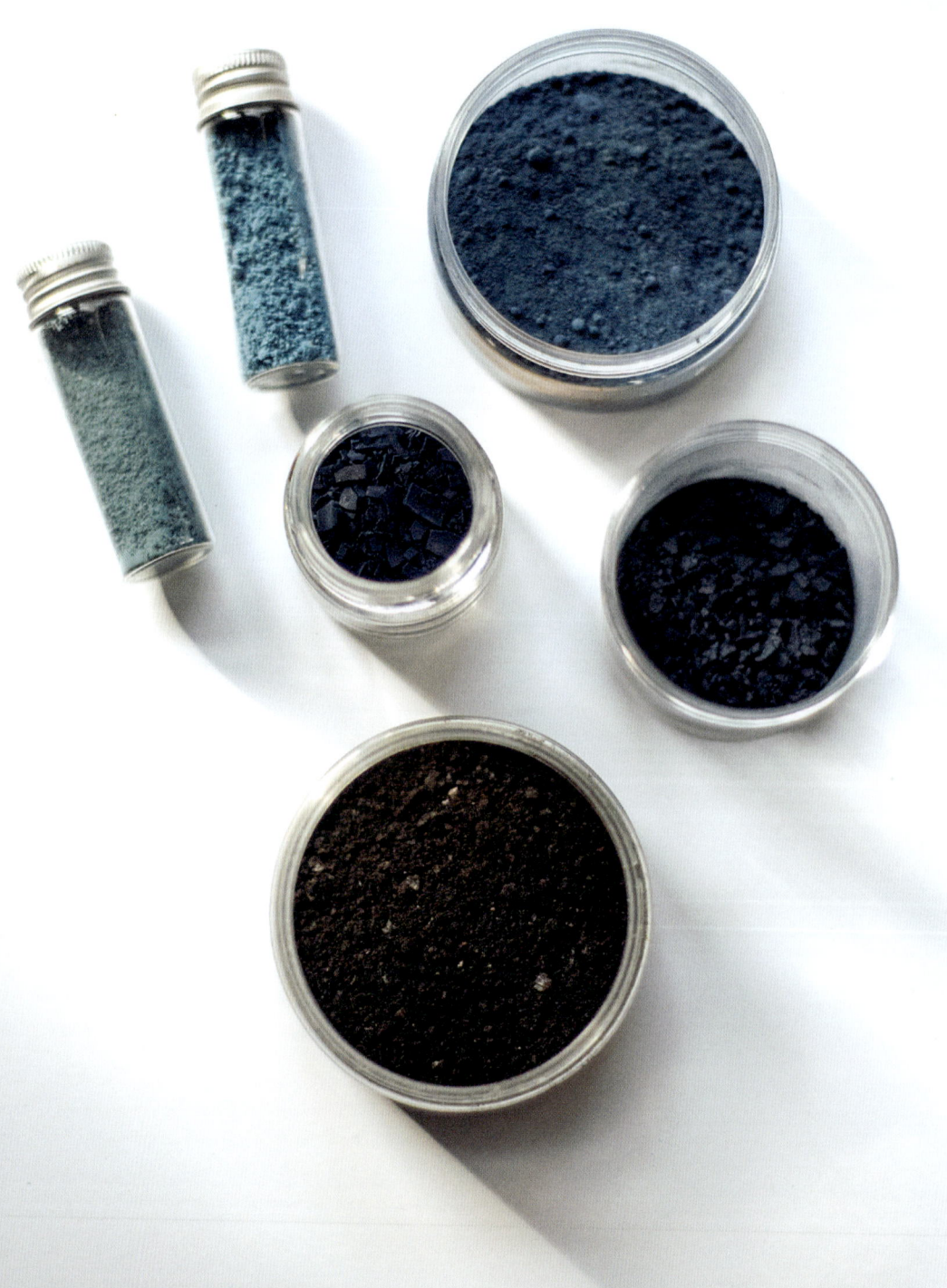

Lakes are one historical technology of botanical-pigment making, but there are a few other methods that expand the rainbow into shades of black and blue. These help fill out the cool end of the palette with lightfast colors—none of which rely on mordants for their backbone.

Black and blue follow different chemical pathways from plant to pigment. Shades of black are made by carbonization: burning plant material to reduce it to a dark essence. Indigo is elicited through fermentation, instigating a cascade of enzymatic and microbial interactions that come to rest at rich midnight blue. And the turquoise shade of Maya blue comes from a unique hybrid nanocomposite of pigment and clay, much more than the sum of its parts.

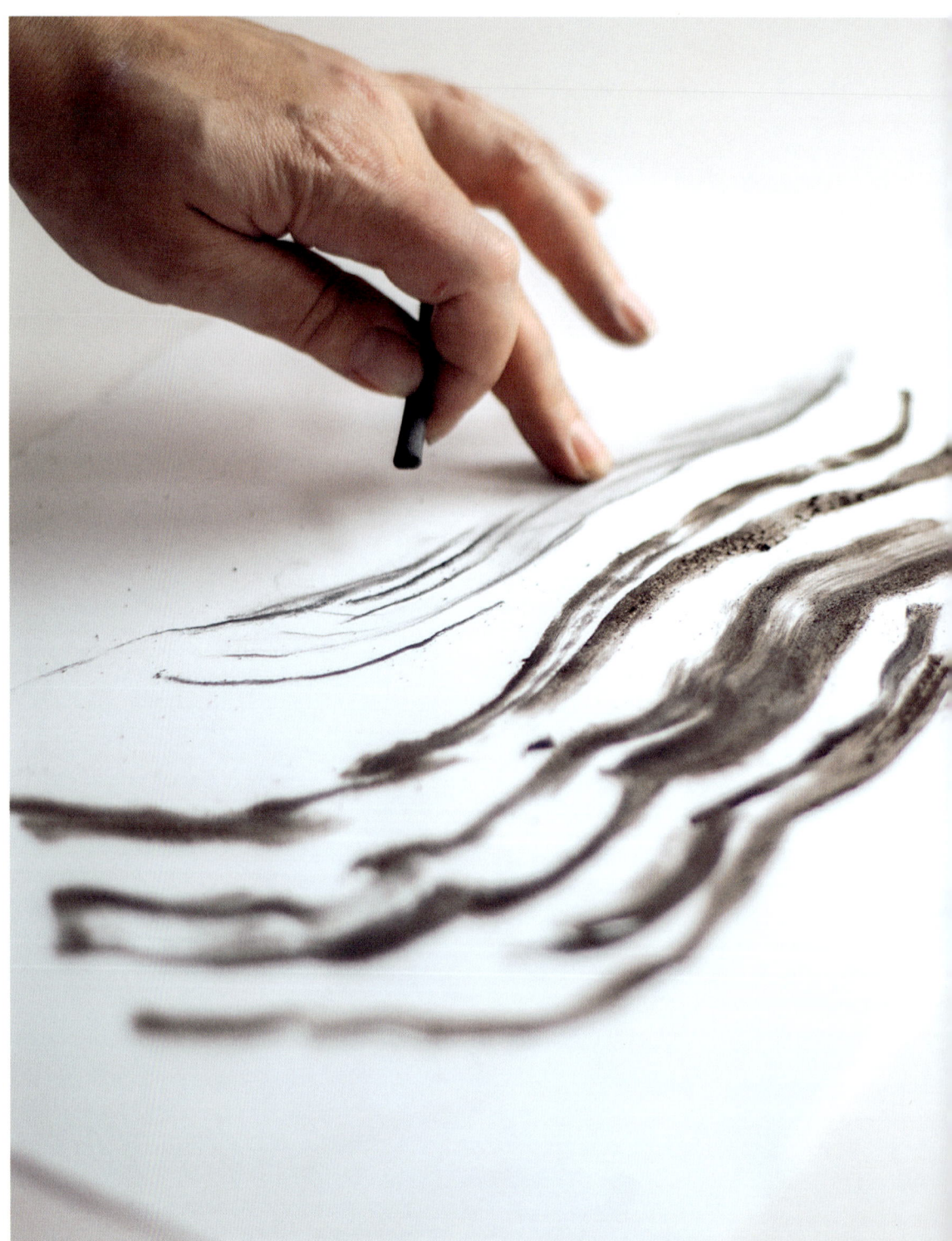

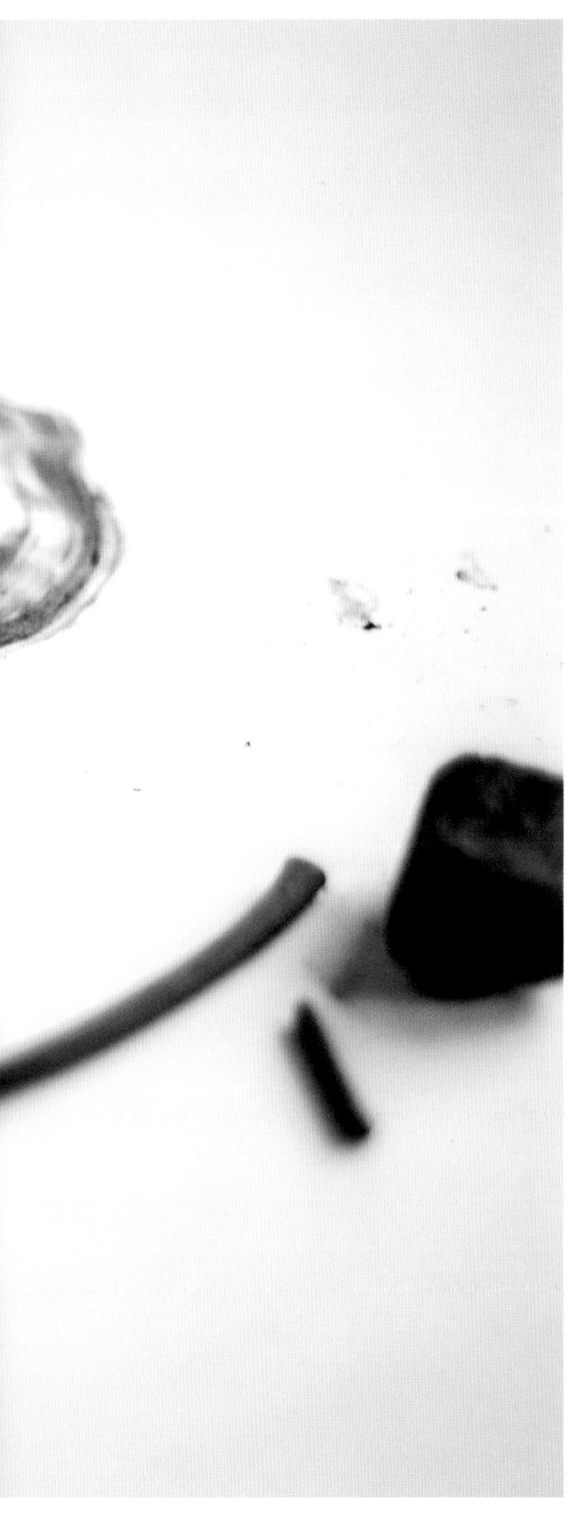

Charcoal

Charcoal is as old as fire and integral to the history of artistic expression. It's a carbon black made by partially burning woody plant material with minimal oxygen. This process of controlled carbonization drives off water vapor and volatile compounds, leaving a rich black. Charcoal is not only a pigment—it is a valuable fuel because it burns at high temperatures. The fields of metallurgy and glassmaking relied historically on specialized charcoal burners: the woodsmen who built giant charcoal kilns in the forest and tended their fires for days on end. (The sooty charcoal burner occasionally appears in folklore as a social outsider who speaks the language of the forest.) Sustainable charcoal production is, in many cultures, woven into traditional ecological knowledge, resource management, and forest stewardship.

The tone and texture of charcoal pigment depends on the type of plant used and the temperature at which it is charred. Fat grapevines and willow twigs make excellent vine black drawing sticks. Wine corks give a soft pigment that can be rubbed directly onto a drawing surface. The pits of peaches, cherries, and similar fruits make a velvety black pigment, but they are too dense to use directly as drawing media. All these varieties of charcoal can be finely ground and mixed with the binder of your choice. They also are completely lightfast.

CHARCOAL PIGMENT RECIPE

I like to make charcoal in an outdoor campfire pit, particularly if I am using a new kiln whose paint will burn off in the process. Choose a time of year with low wildfire risk. Always follow local regulations for outdoor fires and take practical safety measures.

MATERIALS

· Tin canister with a snug lid, such as a tea or biscuit tin

· Plant material suggestions:
    Grapevine or willow twigs (finger-width size), freshly cut
    Fruit stones, kernels, and seeds, such as peach, plum, cherry, grape, or date
    Natural wine corks (check that these are not synthetic or composite material)

METHOD

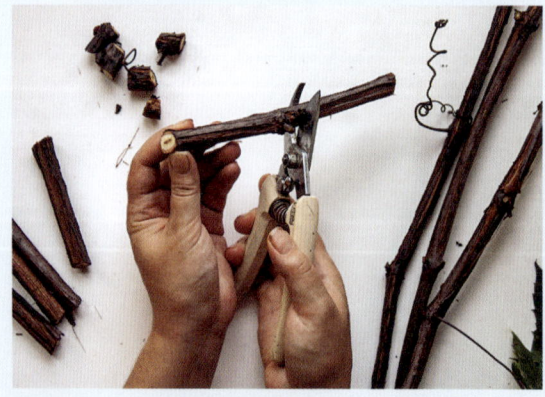

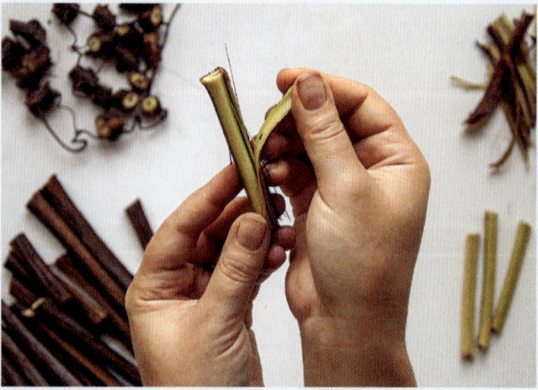

1. Prepare the grapevine or willow by gathering fat vines and branches, preferably more than one year old, during the growing season. Trim off tendrils and side-shoots, and then cut lengths from between the knuckles to fit snugly in your tin.

2. Strip away the bark, down to the pale inner wood. If you are storing branches for future charring, strip the bark while they're fresh, and store the trimmed lengths in a paper bag.

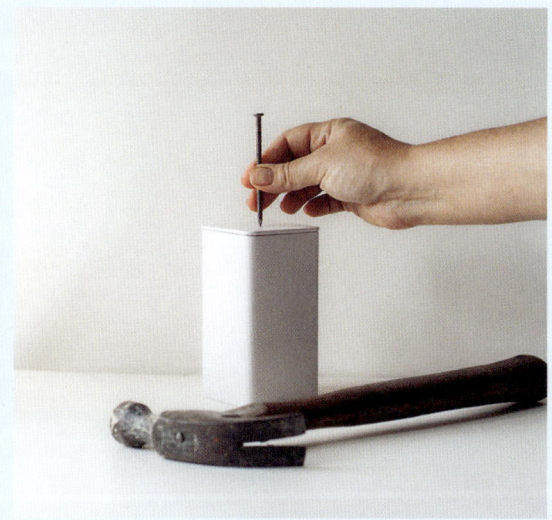

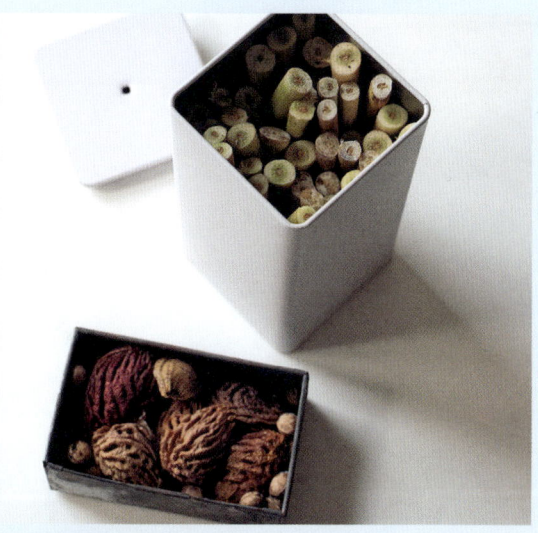

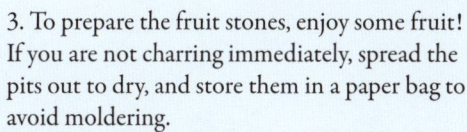

3. To prepare the fruit stones, enjoy some fruit! If you are not charring immediately, spread the pits out to dry, and store them in a paper bag to avoid moldering.

4. The tin will be your mini charcoal kiln and needs a vent to allow gases to escape during heating. Use a hammer and heavy-duty nail to puncture a small vent hole in the lid.

5. Fill the kiln with your prepared plant material, leaving as little air space as possible, and close the lid. To prevent the lid from coming loose, you can wire it closed. Make sure that the wire is not galvanized metal, which will emit dangerous zinc oxide smoke when heated. Alternatively, place a rock on top of the lid before building your fire.

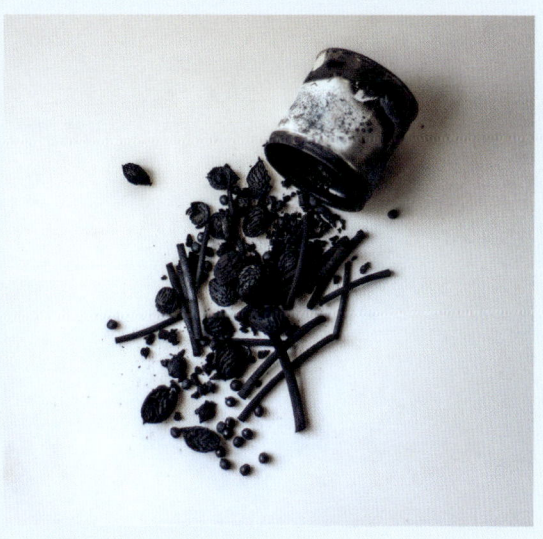

6. Prepare a toasty fire with your kiln at its heart. Once the fire is roaring, allow the charcoal to carbonize for at least one hour. The larger the kiln, the longer it needs to fully blacken. If you see or hear vapor spouting from the vent hole, charring isn't yet complete. Allow the tin to fully cool among the ashes before opening.

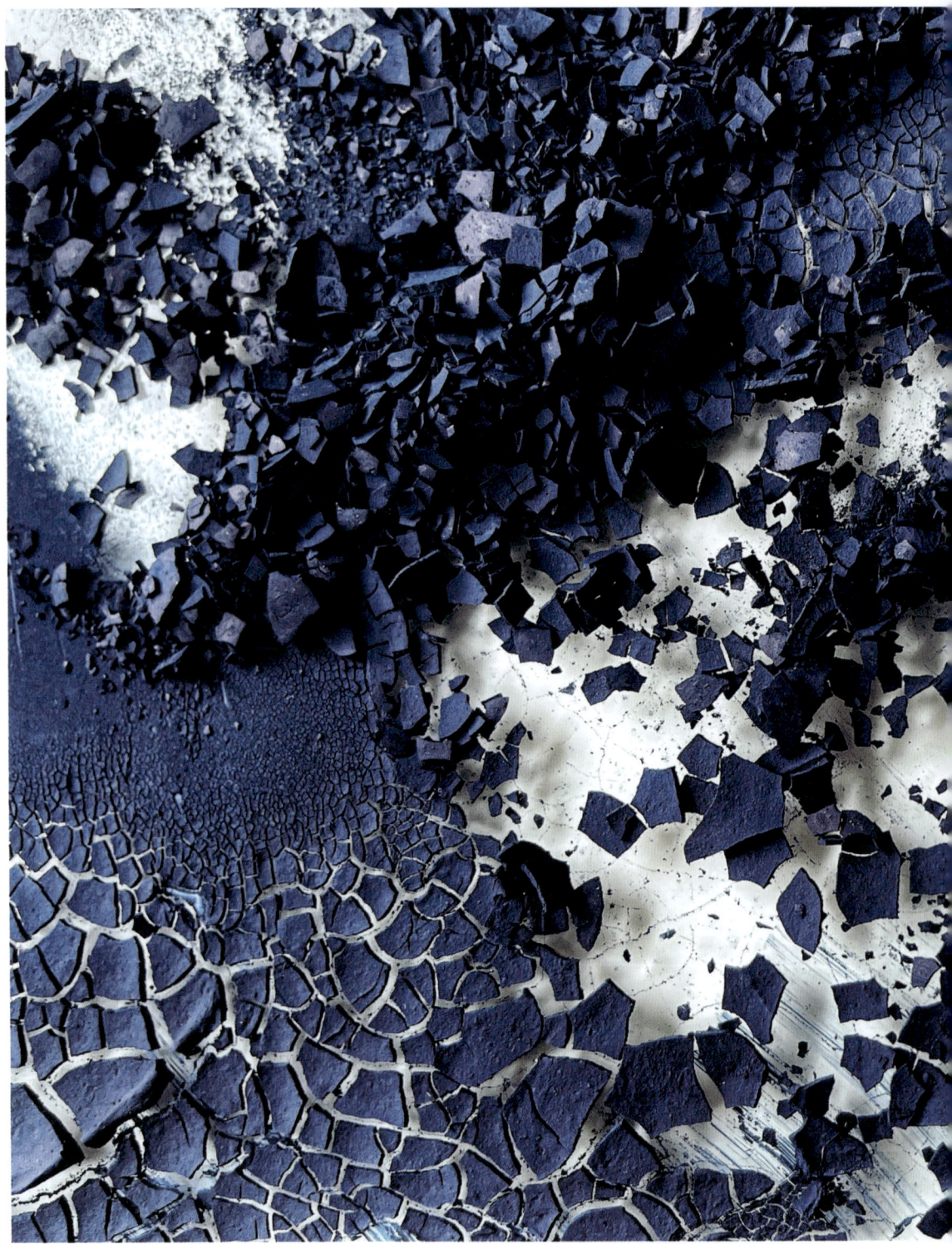

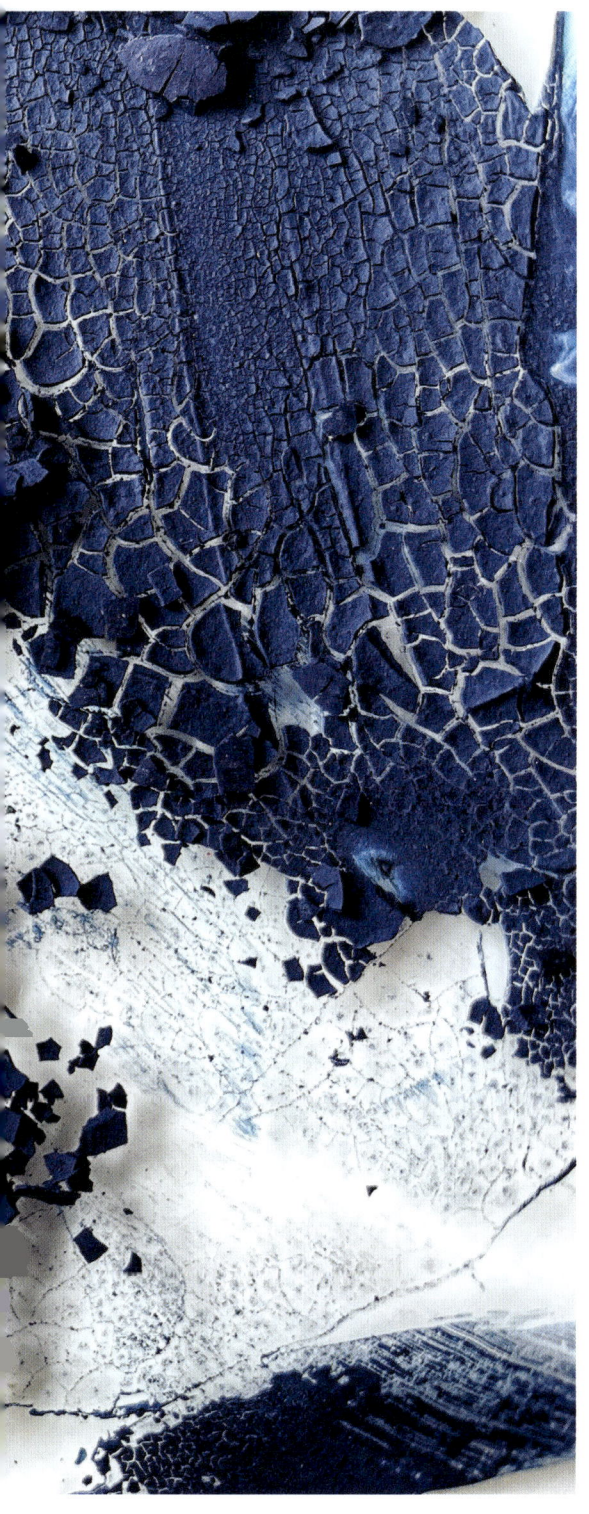

Indigo

Indigo blue never ceases to be a stunning surprise and delight. It manifests seemingly spontaneously from the firmament in the deep azure of the late twilight sky. How could such an ethereal yet enduring color hide within a plant?

Indigo pigment is a product of several unrelated plant species who share the ability to compound its precursor molecules. The blue itself is not present in the plants, merely the potential for blue resides within them. It's wonderful to imagine people from many different cultures over the centuries who developed processes for growing, extracting, stabilizing, and applying indigo blue in both dye and pigment form. Indigo blue has a high lightfastness among plant pigments, and though it is not as fully permanent nor as vibrant as mineral blues, it is much more accessible and renewable.

Unlike the mordant dyes used to make lake pigments, indigo holds no chemical attraction to alum. It is classified as a vat dye, and it follows different chemical pathways into and out of the dye pot—or dye vat, hence the name. Indigo must be chemically coerced into dissolving because the beautiful blue is not soluble: It is an insoluble pigment. Great news for our pigment-making purposes! The blue powder extracted by fermentation is a botanical pigment—no alum, laking, or further chemical tinkering needed. While we will only discuss indigo pigment here, growing and processing indigo might well inspire further exploration of fresh leaf applications and vat dyeing. Go forth! Indigo is as versatile as it is enchanting.

Growing Japanese Indigo

The primary sources of indigo are *Persicaria tinctoria* (Japanese or Chinese indigo), *Indigofera tinctoria* (true indigo), *Indigofera suffruticosa* (Guatemalan indigo), and *Isatis tinctoria* (woad). Depending on the local climate and growing conditions, these indigo-bearing plants can be cultivated by the home gardener, and pigment can be extracted from their leaves using the fermentation extraction method (page 136).

Persicaria tinctoria, which is well-suited to my region, is a straightforward tender annual that grows quickly in a sunny or lightly shaded spot. This riparian plant needs consistent moisture and a nitrogen-rich fertilizer for foliage production, as well as the companionship of growing *en masse* with 15-cm (6-in) spacing. It roots easily if cuttings are placed in a glass of water with a leaf node submerged. Rooted cuttings are a wonderful way to increase your stock of plants or share with friends!

Japanese indigo is ready to harvest when the plants are about 0.6m (2ft) tall and, when bruised, the leaves blush a deep blue color. This generally occurs in late July or August in my garden. Plants will resprout for a second or even third cutting from just a few inches of stem, so leave 10–15cm (4–6in) stubble after harvest, and encourage regrowth with a dose of liquid kelp or aged manure. Japanese indigo is self-fertile, so if you'd like to save seed, leave a few strong plants to flower and set seed at the end of the season.

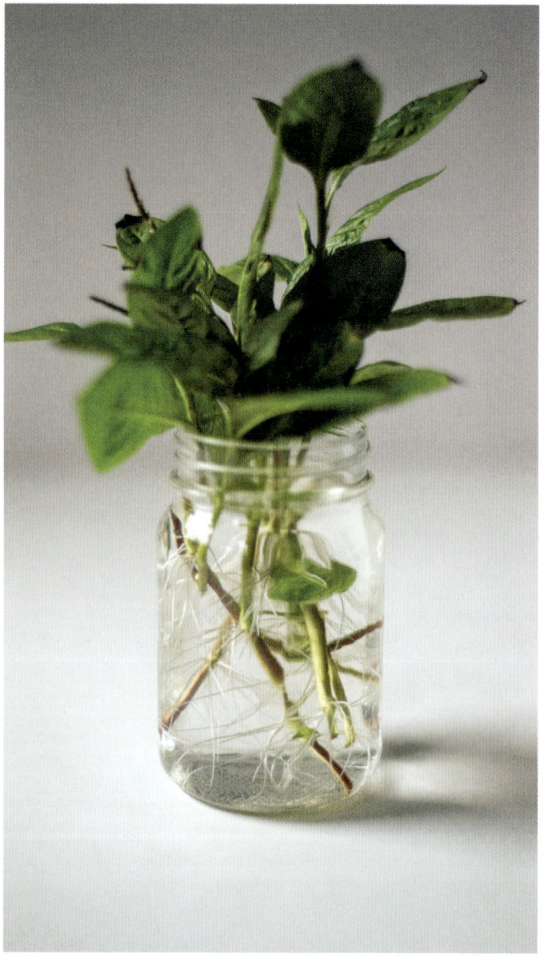

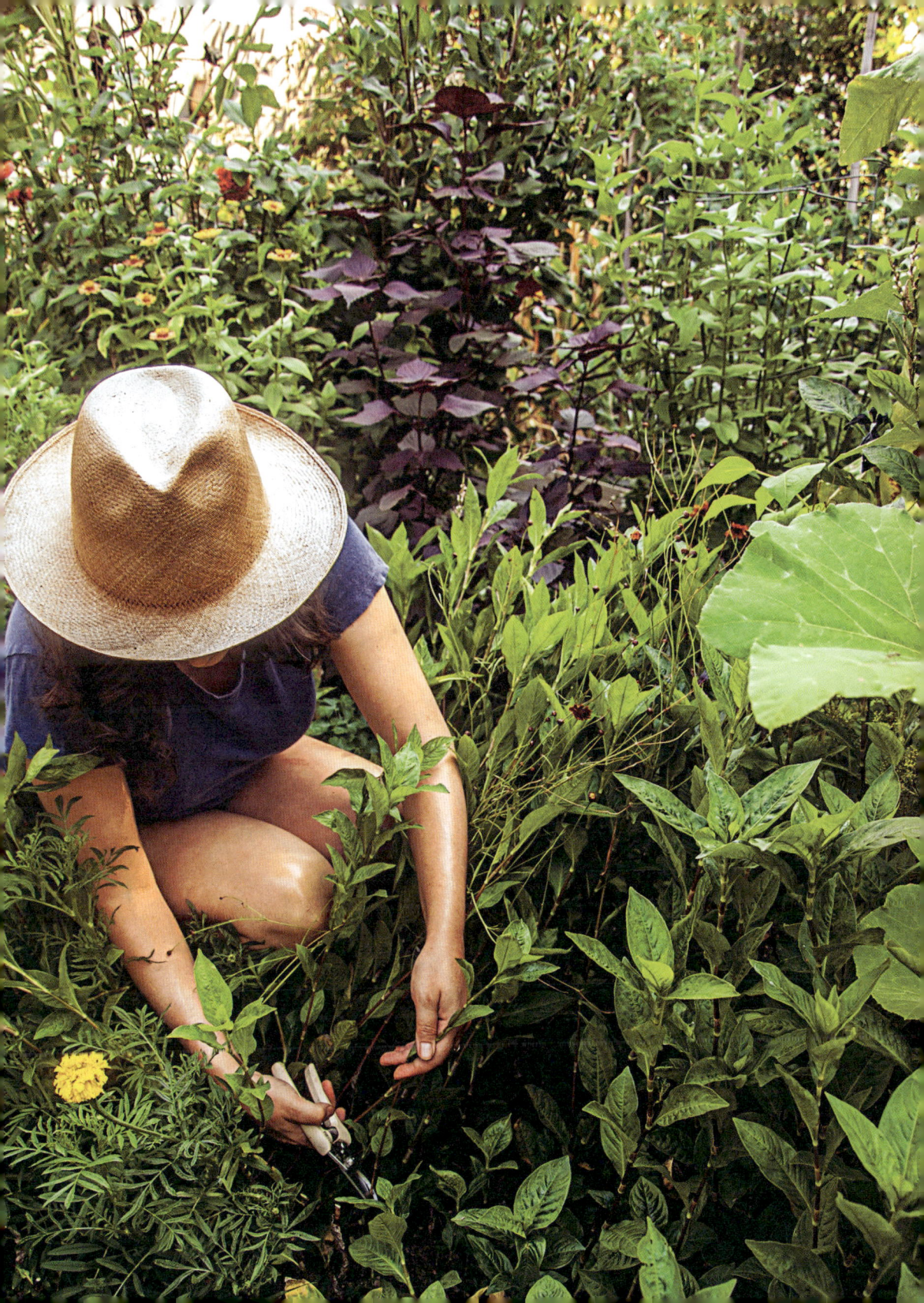

INDIGO FERMENTATION EXTRACTION

Whichever variety of indigo you choose to grow, fermentation of freshly harvested leaves allows naturally present microbes to break down the plant material, releasing indigo's colorless precursor: indican. Enzymes from the leaves, along with the hungry microbes, cleave the indican from its companion glucose, leaving indoxyl: an unstable blue-green intermediary molecule. With the help of alkaline pH and aeration, two soluble indoxyl molecules oxidize and conjoin to form insoluble indigotin: indigo blue.

Though indigo chemistry is complicated, the fermentation extraction has existed far longer than chemical equations. Take the plunge, and after one or two batches of indigo, you will find the process exciting, intuitive, and enthralling. I encourage you to read through the entire recipe before embarking on this fermentation-fueled voyage into blue.

A miniature extraction of 60g fresh leaves in a 1-liter jar is a great place to start. It yields just a scant dusting of pigment, but still offers enough to make Maya blue (page 144). Scale up following the chart below to process a bounteous harvest and increase your yield. For large quantities, I recommend filtering through a 25-micron paint strainer designed to fit in a 20-liter (5-gallon) bucket. Don't forget to support healthy fermentation with a chemical-free water source, such as gathered rainwater.

	FRESH INDIGO LEAVES	ROUGH NUMBER OF PLANTS	FERMENTATION VOLUME	CALCIUM HYDROXIDE*
MINI EXTRACTION	60G	A FEW STALKS	1 LITER (1 QUART)	0.5–1G
MEDIUM EXTRACTION	250G	8 TO 10 STALKS	4 LITER (1 GALLON)	2–4G
I'M-IN-LOVE-WITH-INDIGO EXTRACTION	1KG	1 METER-SQUARE PLANTING	20 LITER (5 GALLON)	10–20G

*Approximate. Many variables are at play!

MATERIALS

· 60g fresh indigo leaves (stripped from a
harvest of approximately 160g plant material)

· 1 liter rainwater, distilled, or chemical-free
water for fermentation

· 1g calcium hydroxide, or as needed

· Distilled water for washing pigment

Because fermentation duration relies on
the ambient temperature, choose a harvest
time when you have a four-day window of
availability and a good weather forecast.
Temperatures between 24 to 35°C (75 to
95°F) encourage the microbes in their work,
but hotter weather can spur the fermentation
to completion in as little as one day (and
easily over-ferment the indigo). Daytime
temperatures of 29°C (84°F) usually result
in a three- to four-day fermentation. Pigment
processing takes about one hour when the
fermentation is ready.

METHOD

Recipe continued on next page

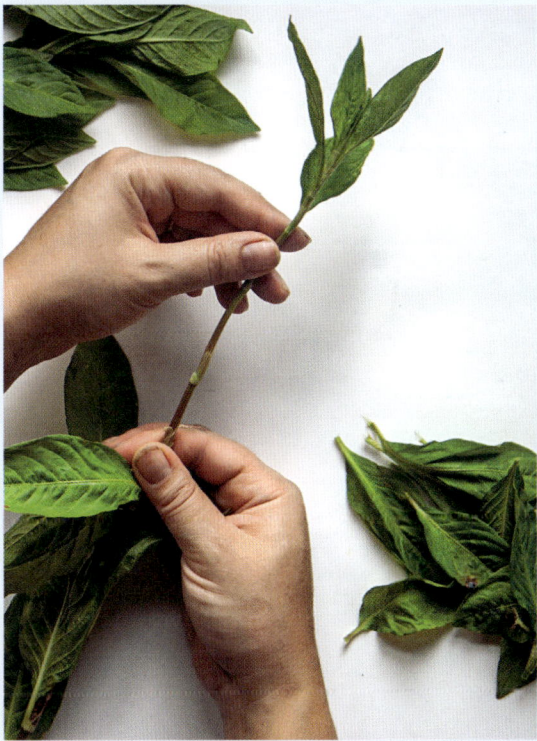

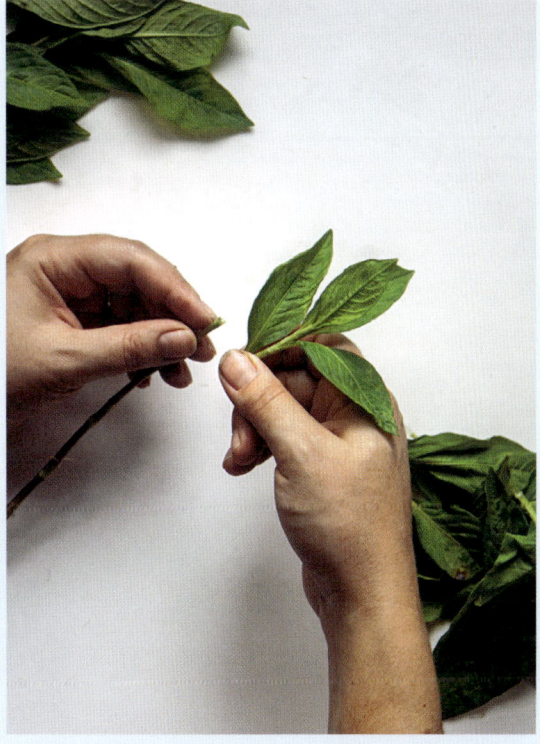

1. Harvest healthy indigo leaves, preferably in the
morning. Strip the leaves from the stems, bruising
them as little as possible.

2. For Japanese indigo, hold the stem near its top,
and run your hand down its length to snap the leaves
crisply off. Pinch the tender growing tip off the stem
and add it to the leaves.

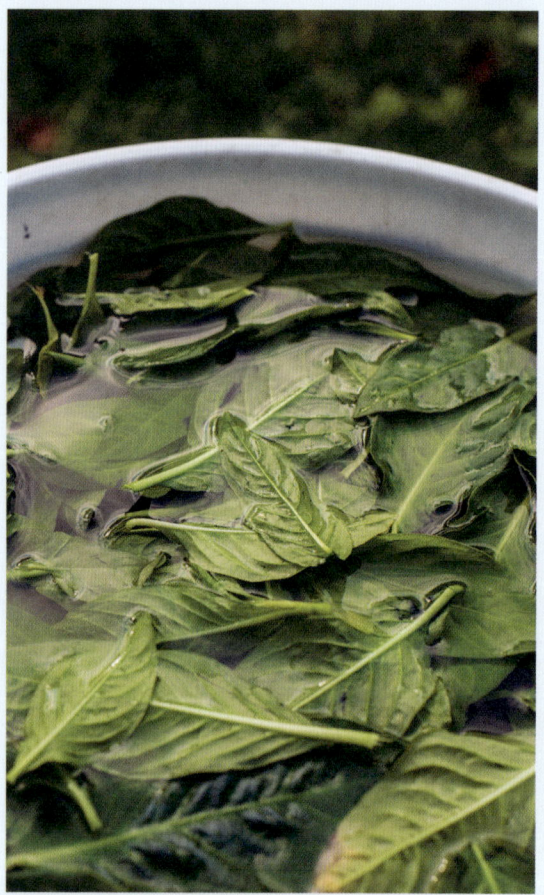

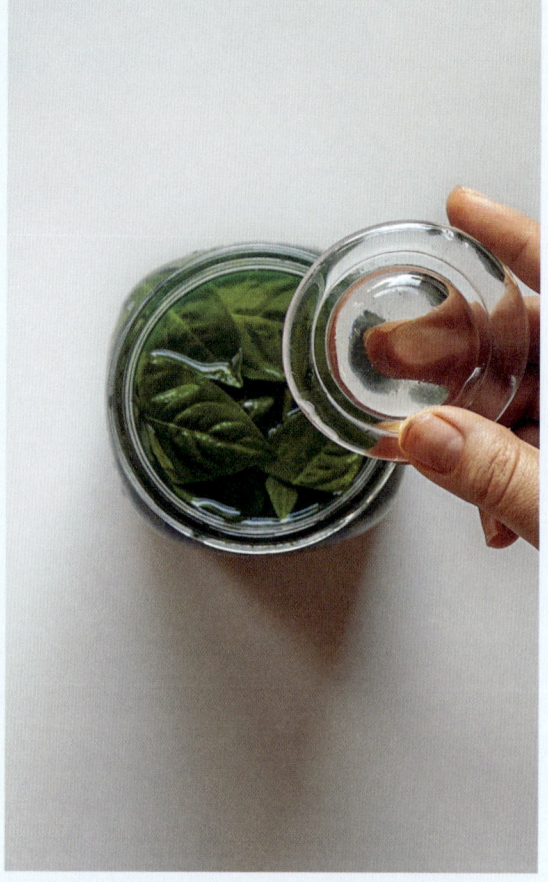

3. If the plants are dirty (say, from rainwater splashing soil onto the leaves), gently waft them through a bucket of cold water to cleanse.

4. Place 60g indigo leaves in a 1-liter container, and cover with water. The leaves should be snug but not compacted. Push the leaves just under the surface of the water, weighing them down if needed with a sieve, small pot lid, or studio-dedicated fermentation weight. Do not seal the fermentation container or stir its contents. Leave the ferment in a warm place (24 to 35°C / 75 to 95°F), where it is safe from curious pets and wildlife. Outdoors is best, as it becomes aromatic over time. If the temperature drops too low, move the container to a greenhouse, place it on a seed germination mat, or tent it with dark plastic to absorb sunlight.

Recipe continued on next page

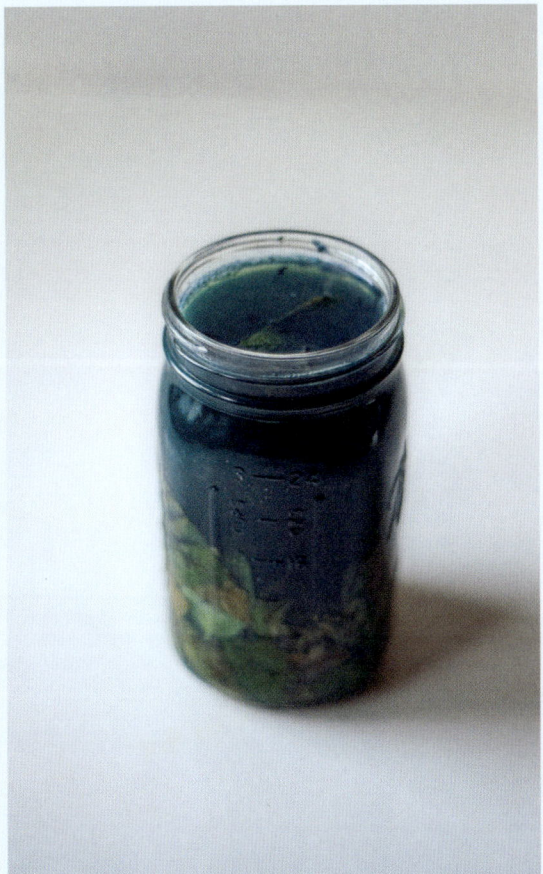

5. Check the ferment two to three times each day, noting the changes to its color and scent. Hold the container to the light or sample its liquid in a small jar to observe its color. Over hours or days, you will see the water first turn yellow (below, top), then green (below, bottom), then a magical opalescent aqua (left). This is colorful evidence of the indoxyl being released. You might notice bubbles rising to the surface, indicating microbial feasting. A crispy metallic sheen might appear on the surface of the water. A fruity, spunky, and briny aroma will develop. The ferment is ripe when you observe some or all these cues.

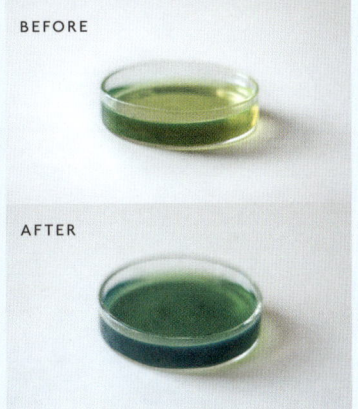

BEFORE

AFTER

PREVENTING OVER-FERMENTATION

If the scent becomes increasingly tinged with an unpleasant rankness, it is teetering toward over-fermentation, at which point the pigment quality declines. If you find yourself in this position, proceed promptly with the extraction recipe. (However, if the ferment grows truly foul—all the leaves wilting to a slimy yellow, the liquid turned brown—then you may compost the lot and start fresh.)

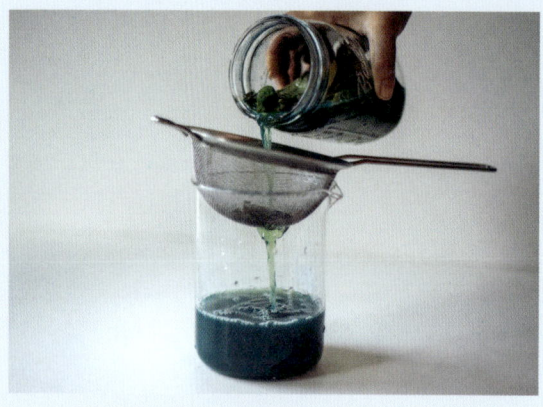

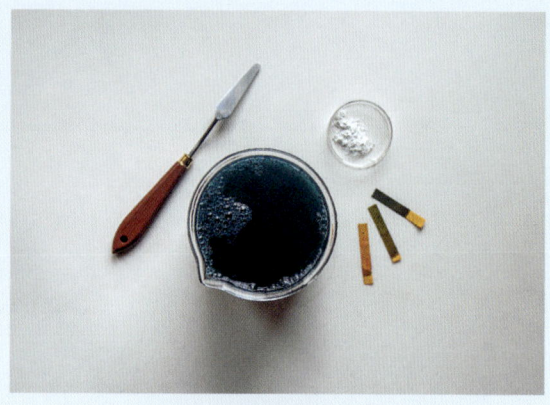

6. Strain out and compost the dyestuff, reserving the aqua-colored fermentation liquid. Move briskly to the next step, as the indoxyl is unstable and quickly degrades.

7. Introduce the calcium hydroxide flocculent: Add 0.25g (about ⅛ tsp) calcium hydroxide to the liquid. Stir, wait five minutes for the powder to dissolve, and test the pH. Repeat as needed until the pH measures 10 to 11, using roughly 0.5 to 1g calcium hydroxide for a 1-liter ferment. Use increments of 1g (about ½ tsp) for a 4-liter ferment or increments of 4g (about ½ tbsp) for a 20-liter ferment.

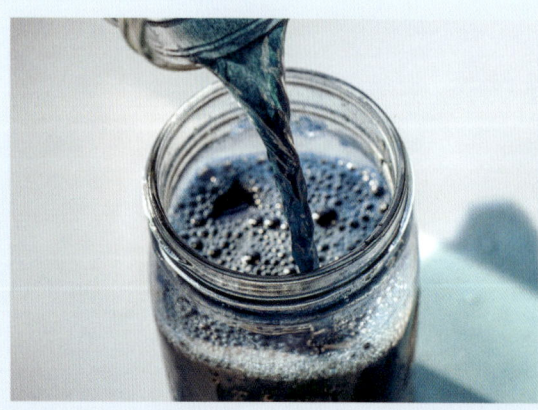

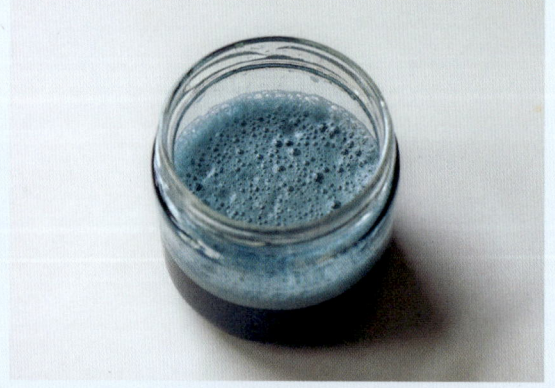

8. Aerate the liquid to oxidize the indoxyl by vigorously whisking or pouring it between two jars. For a large volume of liquid, use an immersion blender or power drill with a paint mixing attachment.

9. You will see the bubbling head of froth turn blue and the liquid go cloudy as the indigo materializes. This froth on the surface of a pigment extraction or dye vat can be used for extra-fine pigment known as flowers of indigo, *aibana*, or *bleu fleurie*. Rest the liquid for 12 to 24 hours, until the indigo pigment settles out.

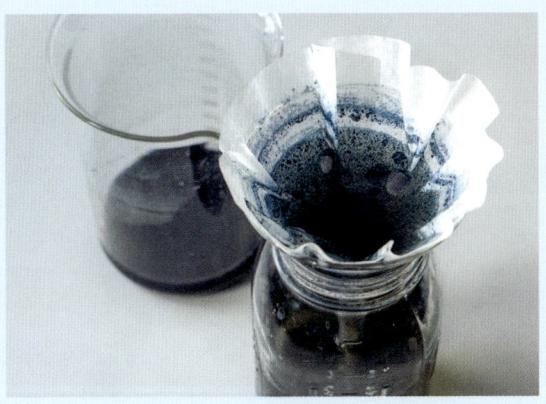

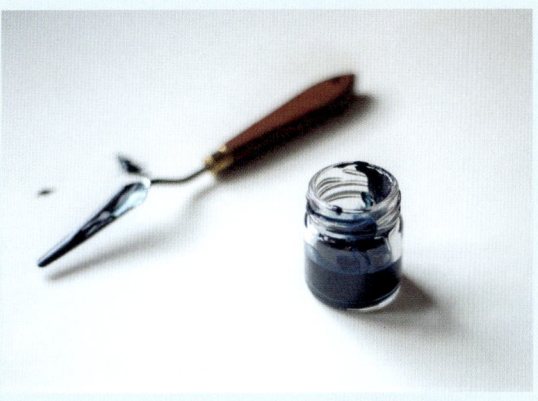

10. Carefully pour off most of the yellow-brown supernatant without disturbing the blue sediment. Filter to separate the indigo pigment from the remaining liquid. To dispose of the supernatant, neutralize the pH with a dash of white vinegar and pour it on your compost or down the drain. Wash the indigo pigment twice following the instructions in the Master Lake Pigment Recipe on page 63, starting at step 14.

11. Optional: For large-scale extractions, store pigment paste in a lidded jar with one drop of clove oil.

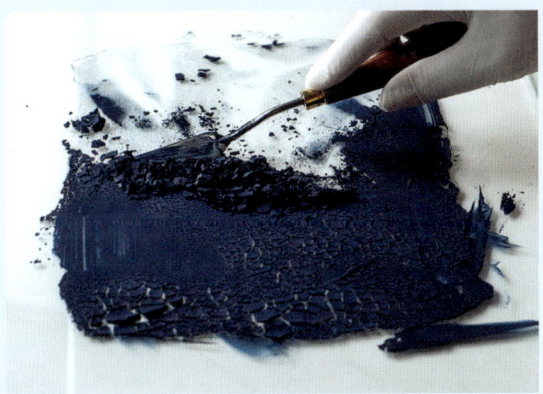

12. Allow to dry to a powder on a filter paper or piece of glass. Then collect the pigment powder with a palette knife and store in a lidded jar.

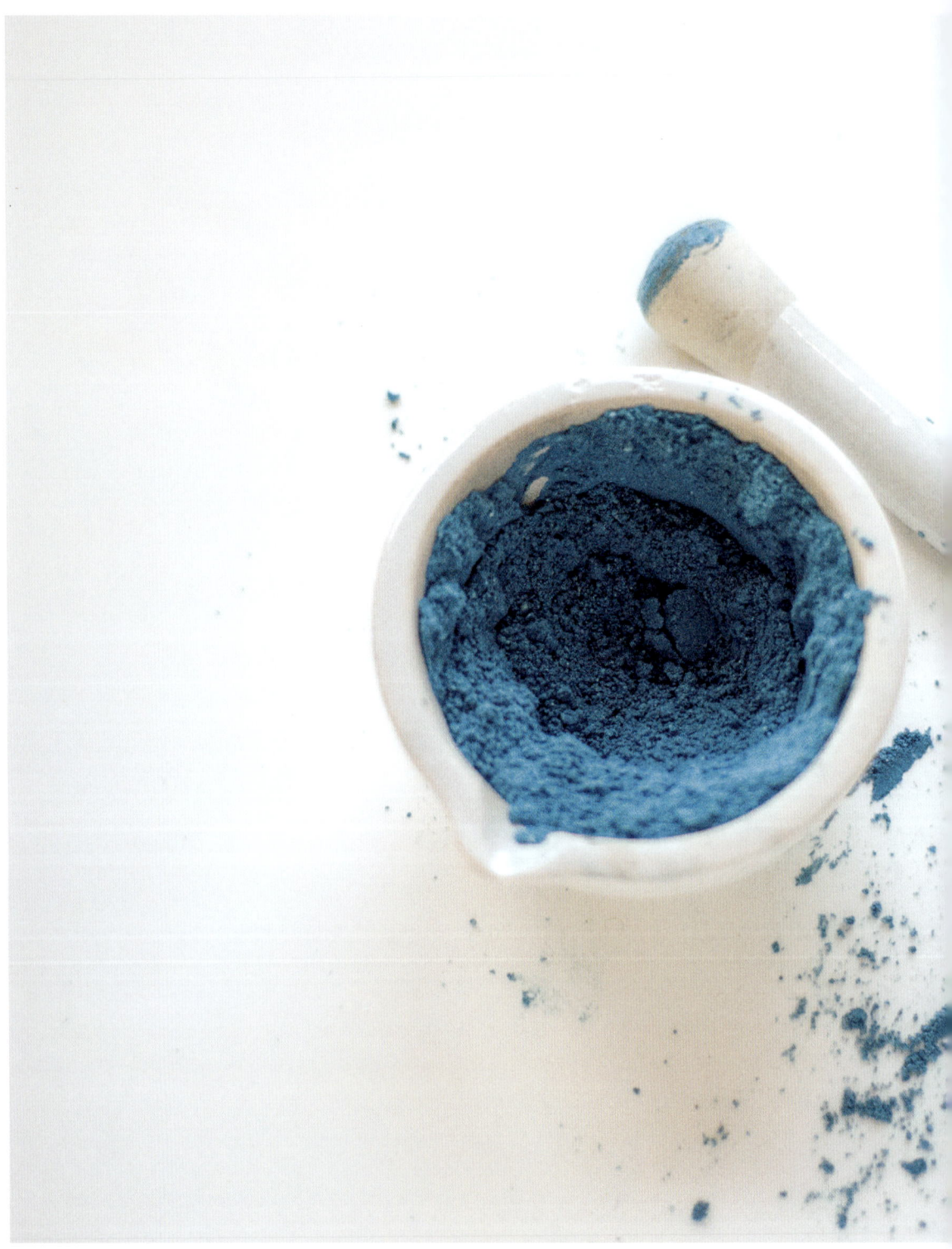

Maya Blue

Maya blue is a luminous range of turquoise-inflected blues made from dry indigo pigment and clay. It was used to beautiful effect on carved statues, ceramics, and murals by pre-conquest Mesoamerican people, including the Maya and Aztec, beginning in roughly 700 CE. By heating the indigo and clay together, the pigment is drawn into miniscule channels in the clay's structure, stretching and stabilizing its color. This means that a small quantity of indigo pigment can create a usable amount of highly lightfast Maya blue. Palygorskite is the type of clay used to make Maya blue historically, but sepiolite (also known as meerschaum) can be more readily available.

MAYA BLUE PIGMENT RECIPE

The source of indigo and the ratio of indigo to clay can provide
a range of shades. A proportion of clay to indigo higher than the
recommended 15:1 will result in a paler shade. I don't suggest using
a lower proportion of clay, because some indigo will be left
unabsorbed by the clay and make for a less lightfast pigment.

MATERIALS

· 1g indigo powder, bone dry

· 15g palygorskite or sepiolite clay

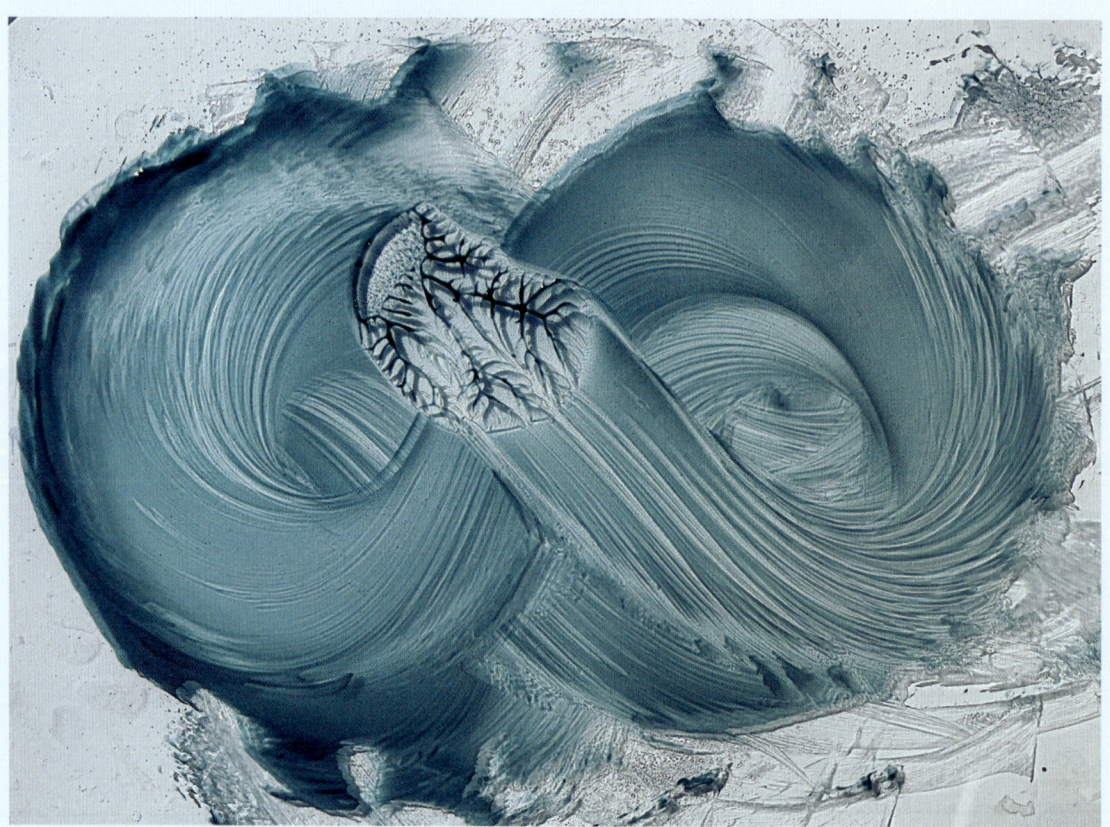

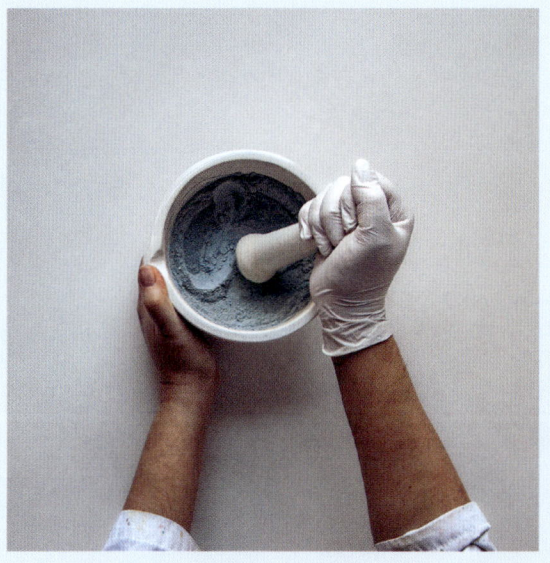

1. Wear respiratory protection and grind the indigo and/or clay in a mortar and pestle until very fine.

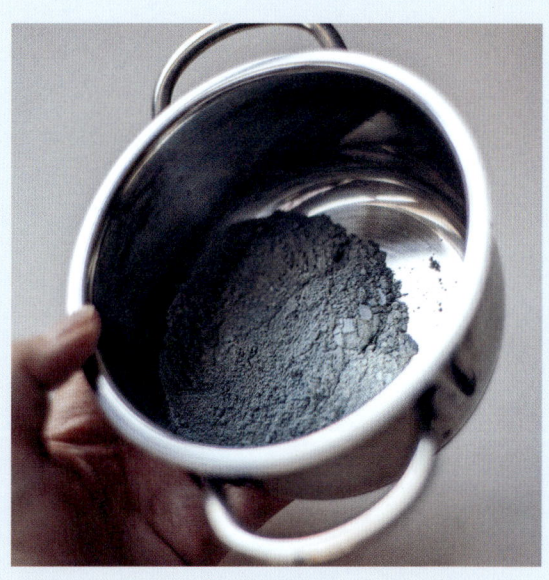

2. Still wearing respiratory protection, combine the indigo and clay in a small, dry saucepan.

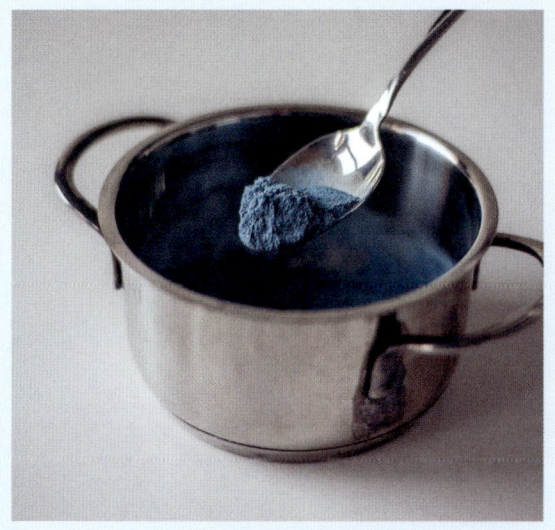

3. Stir over medium-low heat for five to 10 minutes, or until you observe the mixture flush with blue. It should no longer appear to be a drab, heterogeneous blend but rather a unified blue pigment.

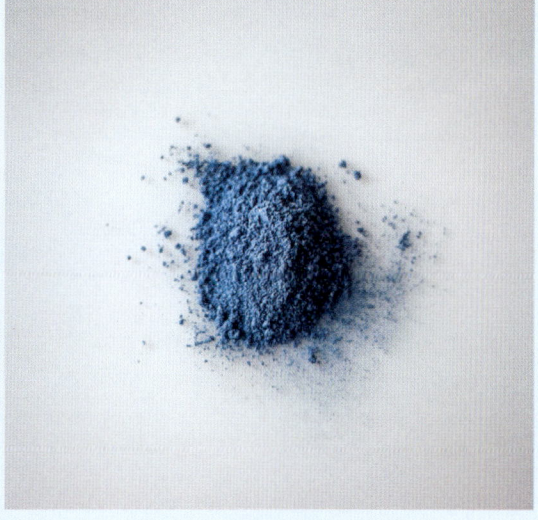

4. Remove from heat and allow to cool. As it cools, the blue's inflection will shift from violet toward turquoise.

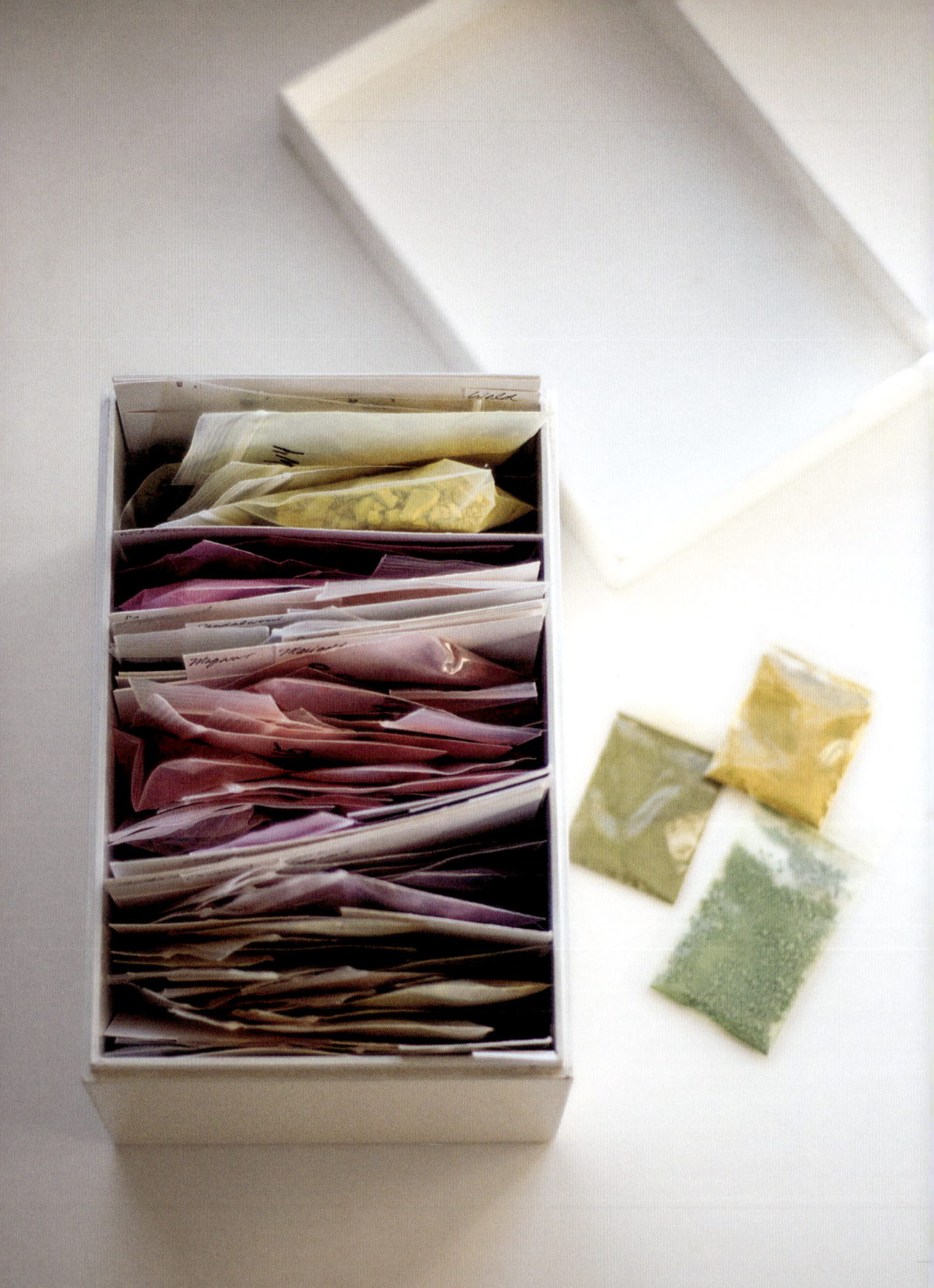

5

ARTISTS' MATERIALS

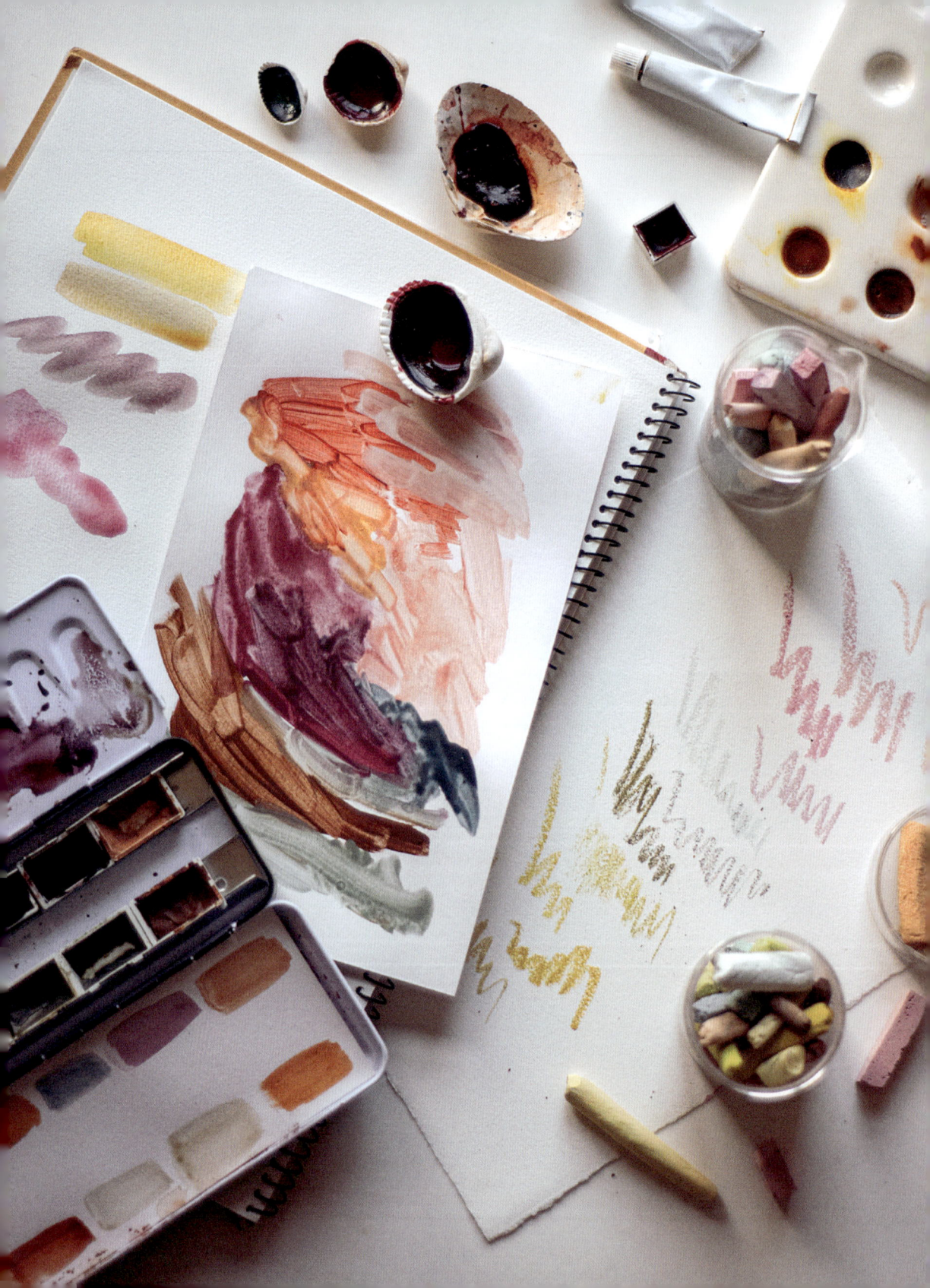

Just as with any earth pigment or synthetic pigment, botanical pigments can be used to make an array of artists' materials. The recipes in this chapter provide approachable methods for utilizing your homemade colors in your artmaking; however, they represent merely a first step into the wealth of possibilities! Homemade art media have more character than their highly engineered commercial counterparts, underscoring the singularity of each color source. Luckily, though the spectrum of lakes is broad, their physical properties are quite uniform. This is because the mineral core of all lake pigments is amorphous aluminum hydroxide, with or without additional chalk. This gives lake pigments predictable handling qualities and means that different colors interact with binders in similar ways. The same is true of charcoal made from different plants, and indigo extracted from different indigo-bearing species.

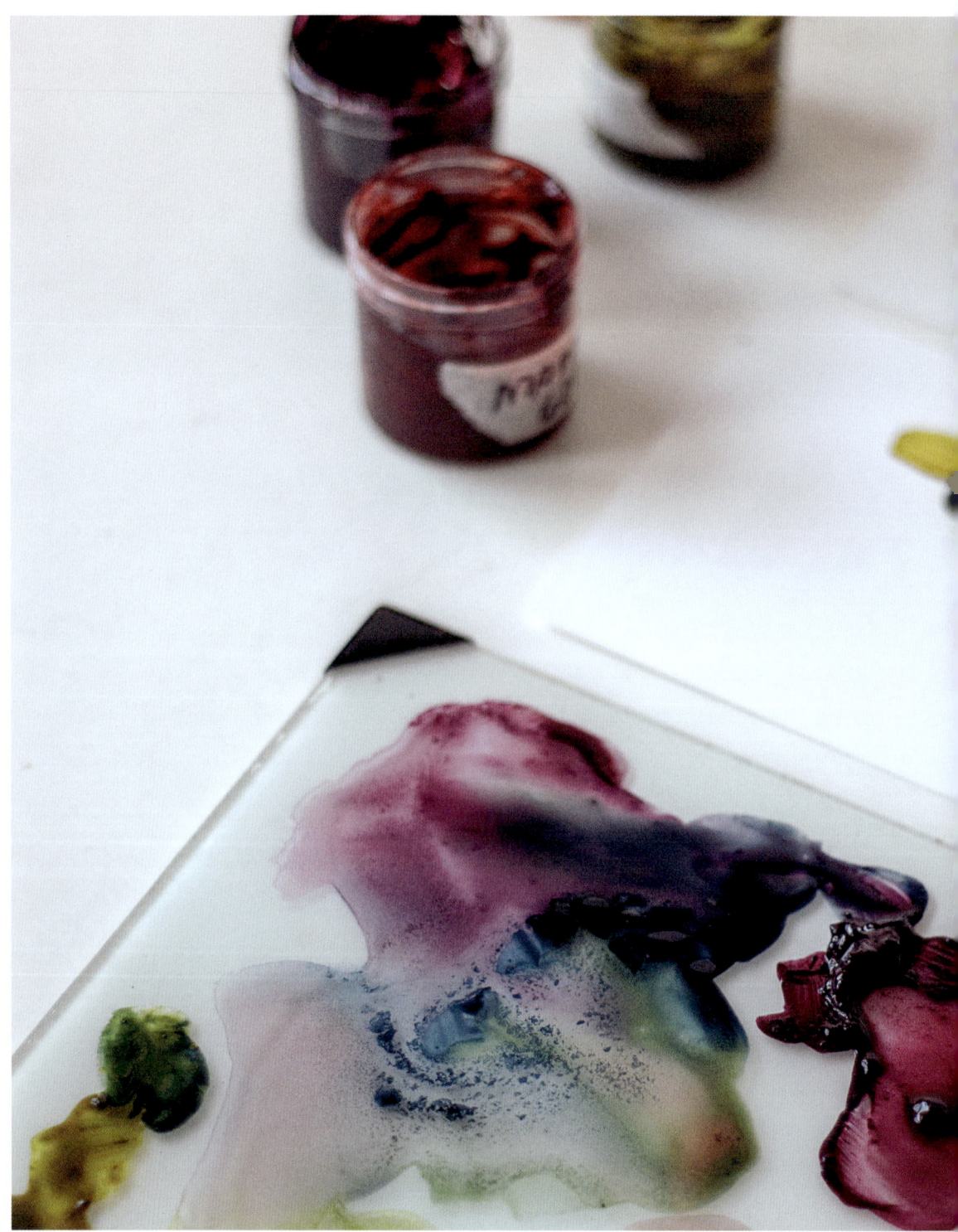

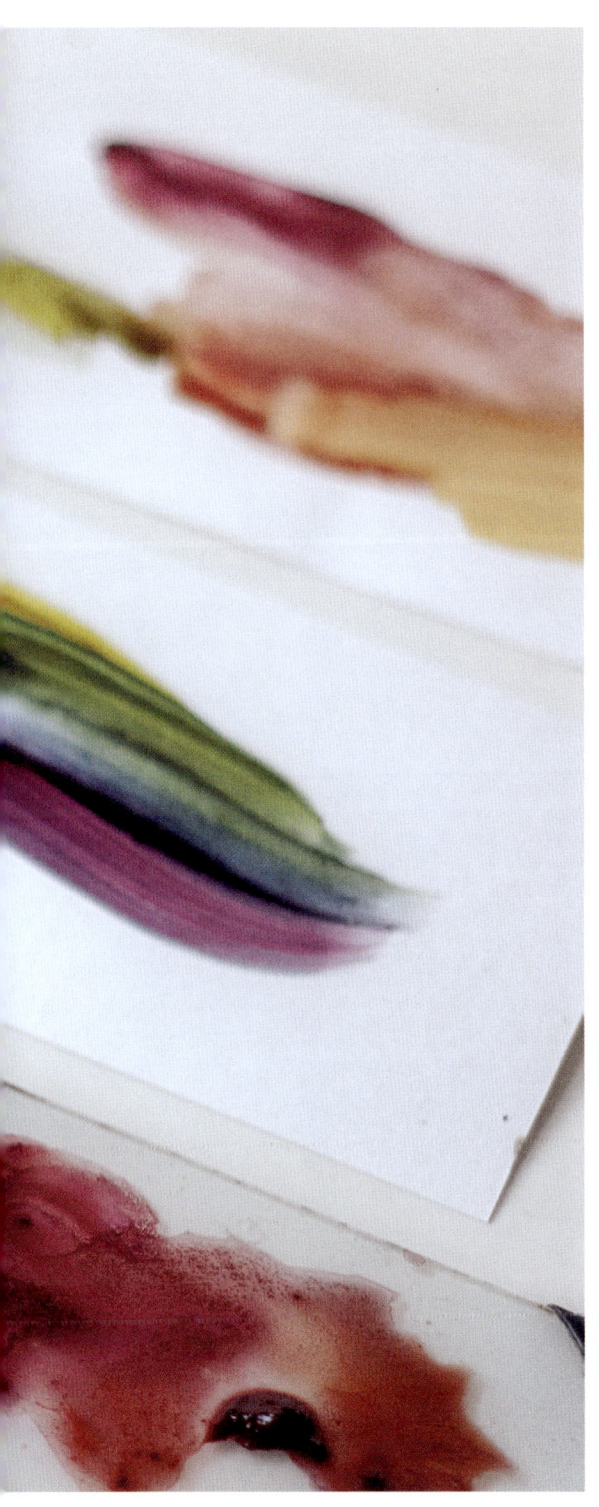

Speedy Paint & Ink

This paint is quick to prepare from washed lake pigment paste or indigo paste, without needing to wait for it to dry to a powder. Because the superfine particulate has not yet agglomerated into clumps, no mortars, pestles, or mullers are required! The silky paste is simply mixed with gum arabic binder. Because the inherent water content of the paste dilutes its color concentration, I recommend selecting strongly hued pigments. The resulting paints maneuver something like tubed watercolors and can be layered in washes. To make free-flowing ink, simply thin with distilled water to the desired consistency. This ink works well with dip pens and paintbrushes, but it is not appropriate for fountain pens.

If you are storing the gum arabic or speedy paint long term, you'll need to prevent their tendency to molder. A single drop of antimicrobial clove oil will act as a preservative, but it can also cause spotty bubbles on the page. I prefer to store jars of gum arabic and speedy paint in the refrigerator in a dedicated box of art materials. This is the necessary trade-off when selecting pigment paste versus powder: the former is not shelf-stable at room temperature, and the latter is.

GUM ARABIC BINDER

You can replace the gum arabic binder with commercially available liquid gum arabic, but it might not be as strong as homemade. Paint out a swatch, allow it to dry, and smudge it with a finger. If the pigment lifts, add more gum arabic to increase the binding strength.

MATERIALS

· 20g gum arabic powder or grains

· 100ml distilled water

· 1 drop clove essential oil (optional)

METHOD

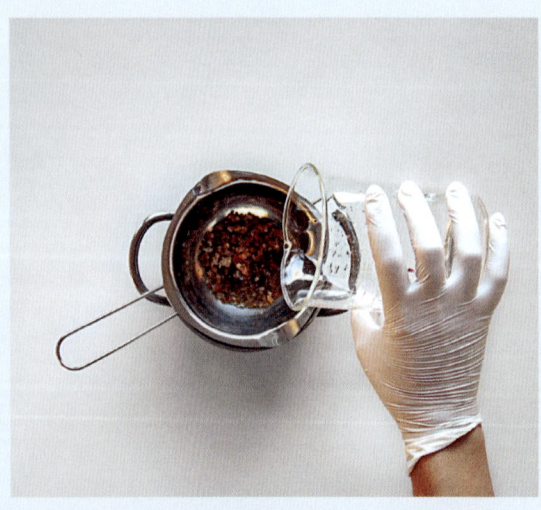

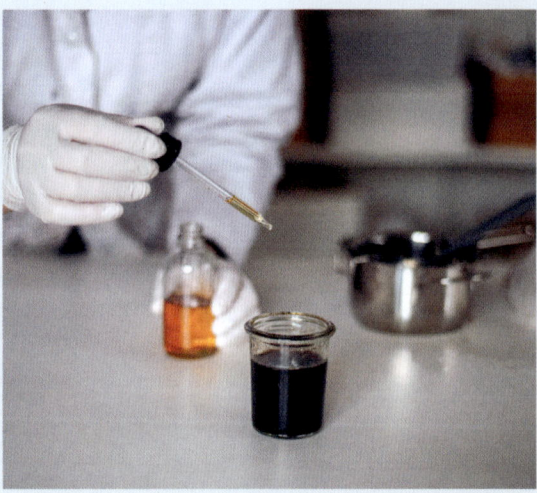

1. Combine gum arabic powder or sap granules with distilled water in the top of a double boiler. Add water to the lower chamber, and simmer for 10 to 20 minutes. Stir the gum arabic mixture in the upper pot occasionally, until dissolved. If needed, sieve to remove grit and twigs.

2. Store in an airtight container in the refrigerator, or add the optional clove oil and store at room temperature.

SPEEDY PAINT & INK

The speedy paint method works just as well with other water-based binders, including egg tempera, glair, and soya milk. Unlike gum arabic which can be refrigerated, none of these binders stores well, so mix up only the amount of paint you can use in one sitting.

MATERIALS

· 1 tbsp (about 15ml) lake pigment paste or indigo paste, washed and filtered through steps 14–16 of the Master Lake Pigment recipe (pages 66–67)

· 1 tsp (about 5ml) gum arabic binder

· Distilled water, to achieve desired flow (optional)

METHOD

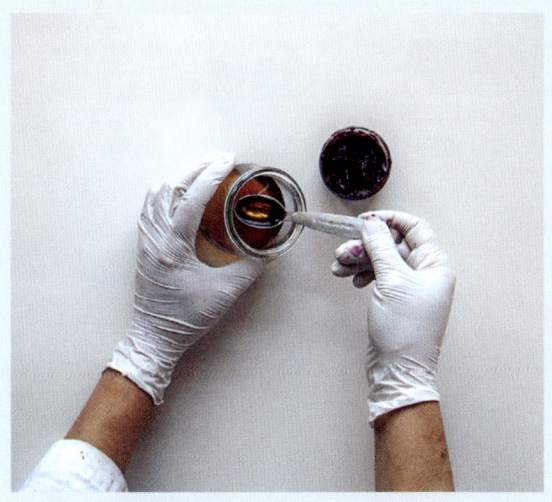

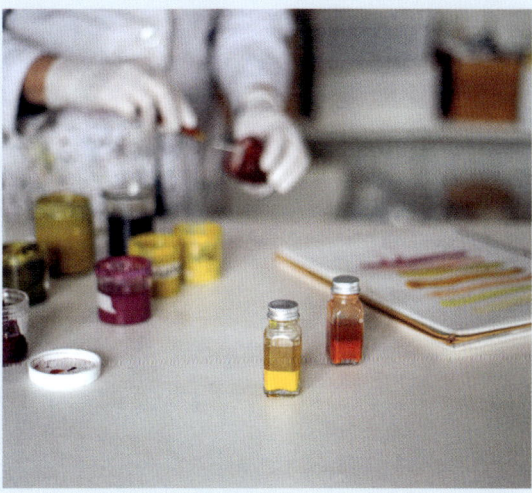

1. Stir 1 tsp gum arabic binder into 1 tbsp lake pigment paste or indigo paste. Multiple colors can be combined, if you like. Optional: Stir in distilled water to achieve desired flow.

2. Store paint or ink in a sealed container, and do not allow it to dry out. If you have not opted to add clove oil to the binder, store paints in the refrigerator. Before use, stir or shake vigorously to draw the settled pigment into suspension.

Processing Dry Pigments

The texture of lake pigments varies considerably with the factors of their synthesis: dye source, temperature, volume of water, base, order and speed of addition, and drying conditions. As the wet pigment paste loses moisture, it fuses into a mosaic of chunks through the process of agglomeration. These clumps must be manually ground into a powder, separating the agglomerates back into minute particles. Well-ground pigment makes for smooth paint and pastels. I pass pigments through two phases of grinding, beginning with this rough processing in a mortar and pestle. Then, I determine if the pigment will be used for dry or wet media. For dry media, including pastels, I use a stacking sieve to sort the fine pigment from the coarse pigment (page 157). For wet media, including paint and ink, I use a muller and slab (page 164).

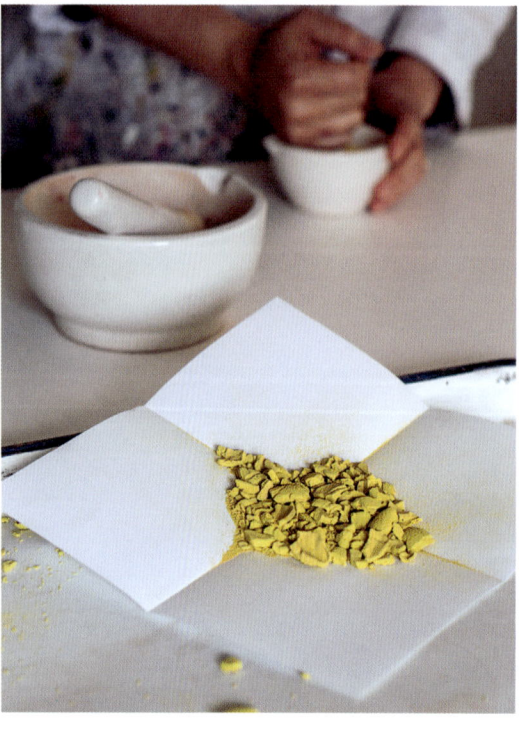

The ultrafine dust thrown up by the grinding process is a respiratory and eye irritant, and can cause immediate discomfort and long-term health issues with repeated exposure. Always protect yourself while handling and grinding dry pigments. This includes not just your airways, but also the clothing and studio surfaces that can become a reservoir for pigment dust.

· Wear a snug, high-quality face mask or respirator, gloves, and eye protection.

· Wear a duster or coverall over your clothes. (Immediately launder your garments afterward.)

· Work outdoors, if possible, with the breeze at your back. Alternatively, invest in a studio ventilator with a flexible exhaust pipe to suck the air from your workstation.

· Work on a nonporous surface and wipe it down when finished.

· Do not sieve pigments in the open air; use a covered stacking sieve to grade pigments if needed.

GRINDING PIGMENTS

This preliminary phase of pigment processing refines the coarse agglomerates to a smooth powder. Take the time to make it as fine as possible, in order to ease the next phase of processing for dry or wet media.

MATERIALS

· Dry pigment

· Tray or large sheet of paper

· Mortar and pestle, approximately 150ml capacity. Scale up to 600 to 800ml for large batches.

METHOD

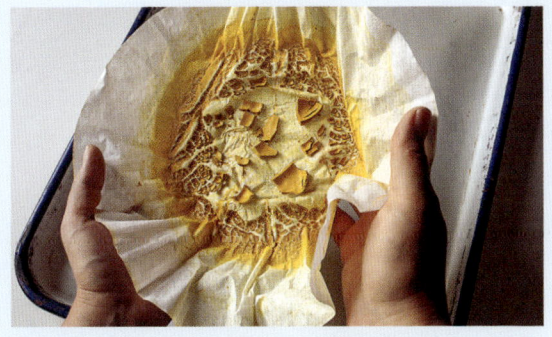

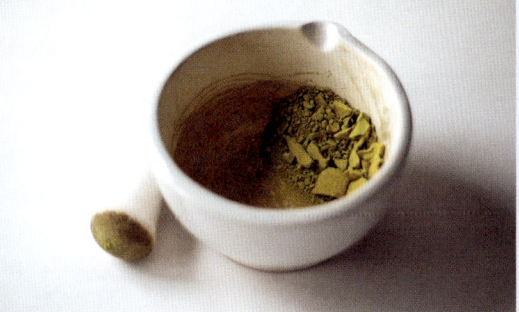

1. Follow the safety guidelines listed to protect yourself from pigment dust. Hold the filter paper of bone-dry pigment over a tray or large sheet of paper to catch pigment. Fold the filter paper and gently rub the halves together to dislodge pigment. Work your way around the filter until it is as clean as possible.

2. Transfer the pigment chunks to a mortar on a sturdy work surface. Use the pestle to gradually grind it to a fine powder. Lake pigments made with chalk and those dried without additional heat are generally easier to grind. Rub a bit of pigment between your finger and thumb. If it feels gritty, continue grinding. If its grains are nearly impalpable, continue to the next phase.

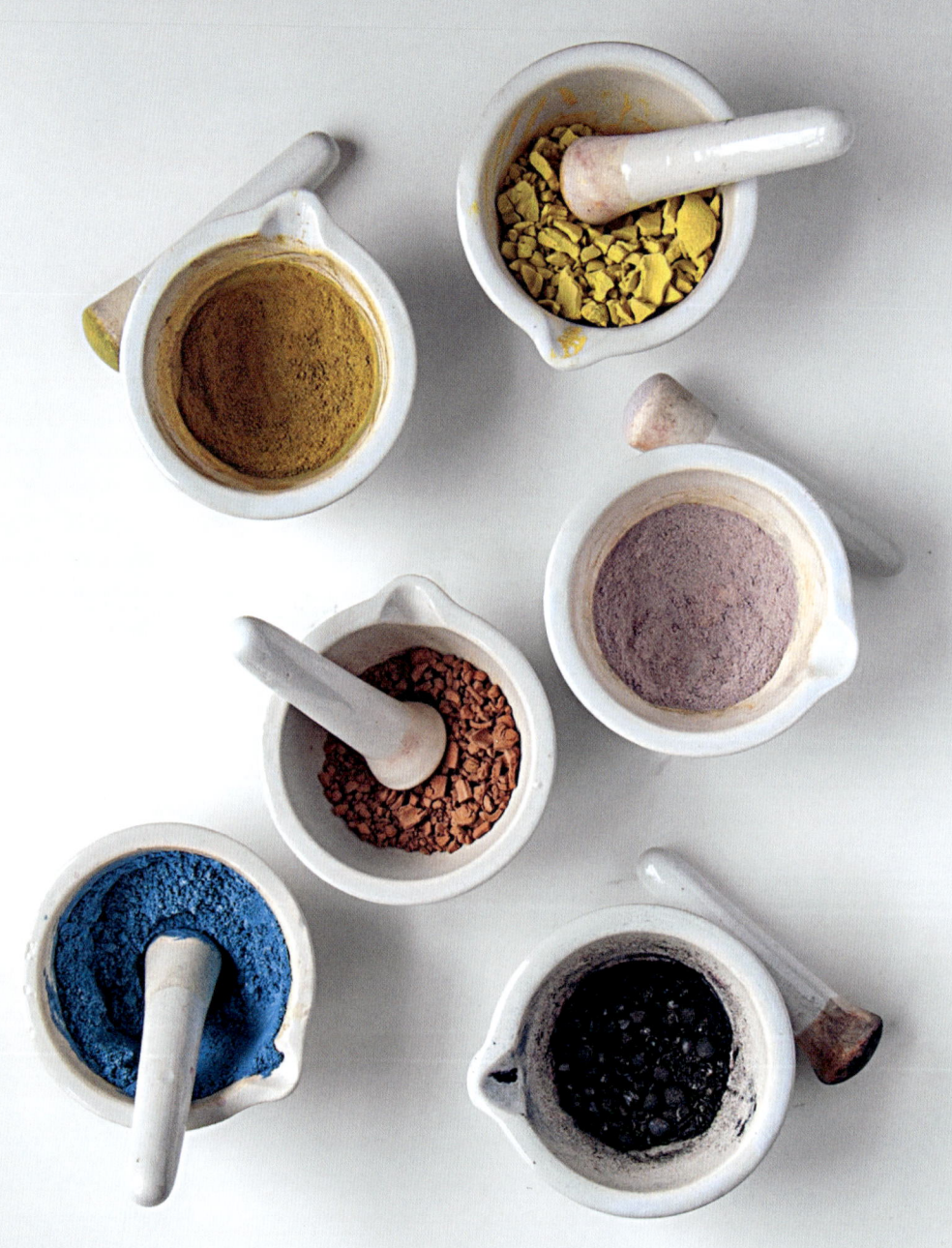

SIEVING PIGMENTS

Pastel sticks, wax crayons, and encaustics each incorporate dry powdered pigment into a binder before setting firm. It's not practical to grind the pigment and binder together, so the pigment must be finely processed first. A stacking sieve is useful for grading pigment by particle size while keeping dust enclosed. The sieve that I use has four levels with meshes between 60 and 325.

MATERIALS

· Dry pigment, ground through step 2 on page 155

· Stout stiff-bristled brush

· Graded stacking sieve

METHOD

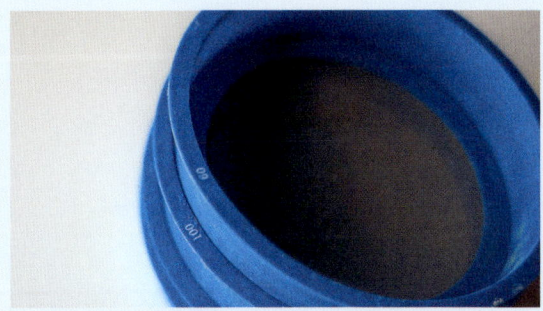

1. Stack your sieves with a lid on top, sieves ordered by increasing fineness, and a catchment at the bottom. Higher mesh numbers correlate to finer screens.

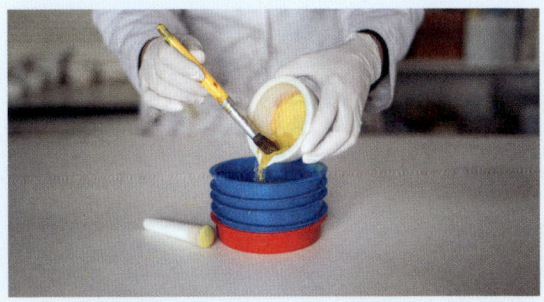

2. Use a stiff-bristled brush to transfer ground pigment from the mortar to the top level of the stacking sieve. Place the cover, and shake the sieve or tap it against your work surface for four to five minutes. Set the sieve aside for a few minutes before opening, allowing the dust inside to settle.

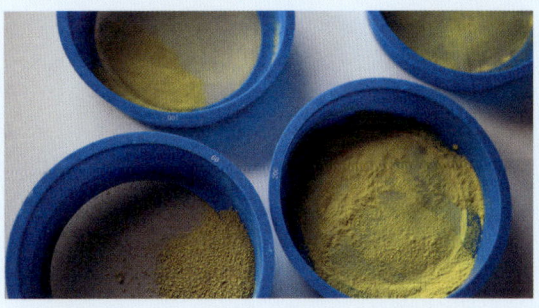

3. Wearing respiratory protection, open the sieve and check your progress. Return coarse pigment agglomerates to the mortar for further grinding. Fine pigment sieved through mesh 200 or higher can be set aside for pastel (page 159) or other dry media.

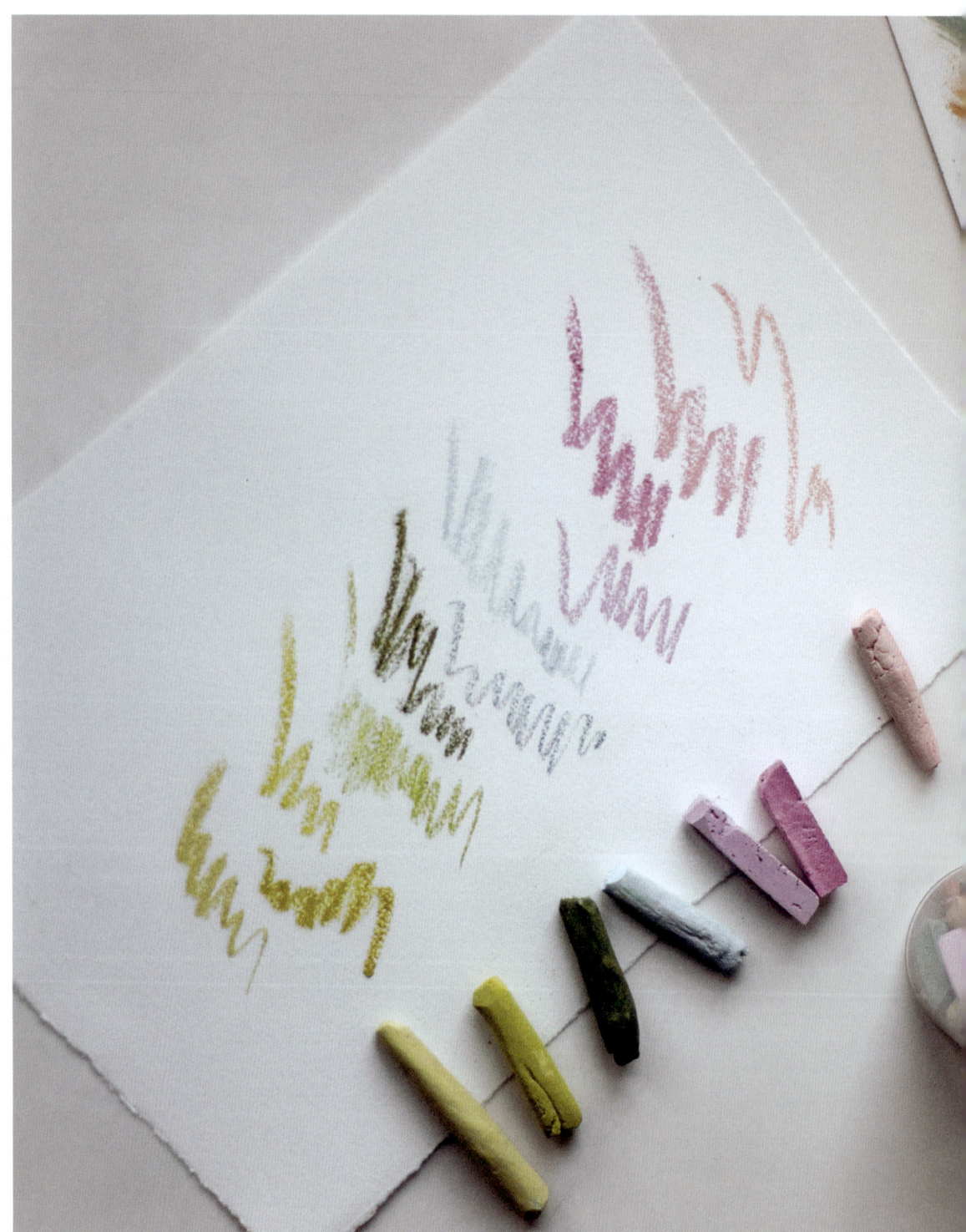

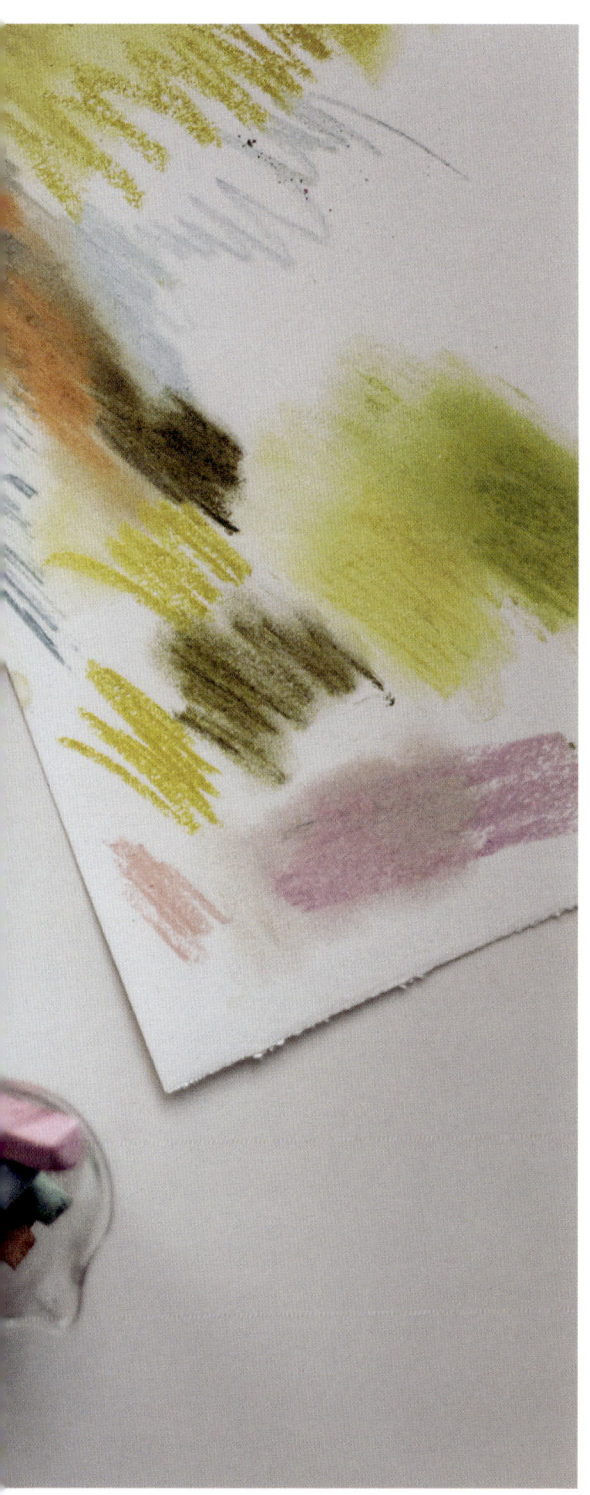

Pastels

The original pastels were soft rocks or colored chalks, rubbed onto a textured surface. Today's pastels are made from pure pigment with a weak binder, rolled or molded into drawing sticks. Applied to a toothy or abrasive paper, the pigment is sheared into the surface and clings loosely in place.

Pastels made from botanical pigments don't strictly require (but can benefit from) the addition of chalk, which gives them a creamy texture that melts into the page. Maya blue responds particularly well to the addition of an equal measure of chalk. The exception is chalk-based lakes, whose texture is, of course, innately chalky. Each small pastel stick requires approximately 2 tsp pigment in total. Mixing pigments together, then shading and tinting the hues with charcoal and chalk, creates a broad palette of pastel sticks.

PASTEL STICKS

These pastels call for a simple binder made from oats, rather than the more traditional gum tragacanth.

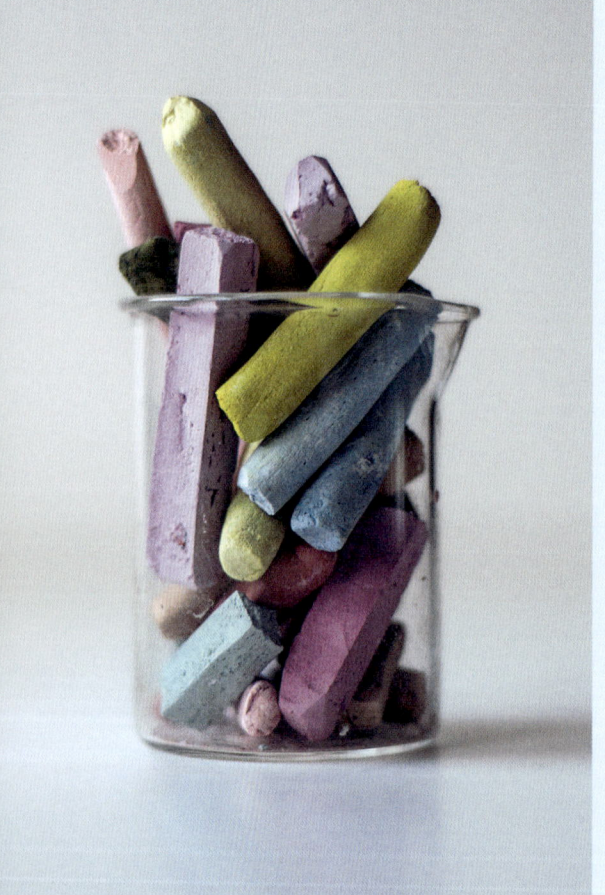

MATERIALS

· 7.5g (1 tbsp) oats (I've had success with old-fashioned and quick rolled)

· 240ml (1 cup) water

· Pigment, finely ground and sieved (page 155–157)

· Chalk, optional

· Newsprint or newspaper

METHOD

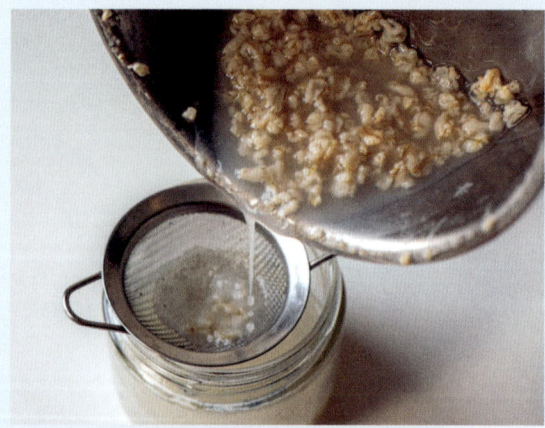

1. Combine oats and water in a small saucepan. Bring to a vigorous simmer for five minutes. Keep a close eye on the oats as they cook; in my experience, oats are jealous of attention and boil over into a sticky mess the instant I look away! Sieve through a fine mesh, and discard or compost the oats. Allow the oat water to cool. If necessary, store for up to four days in the refrigerator.

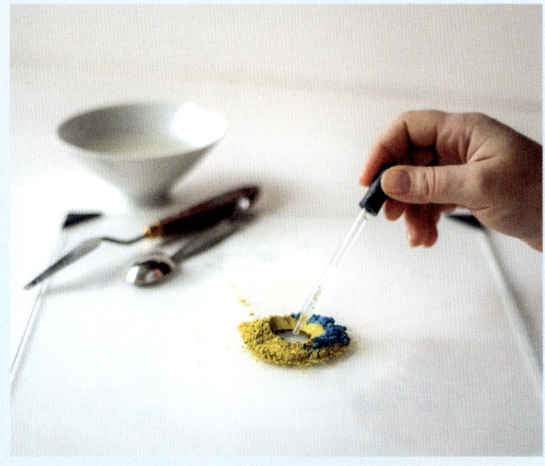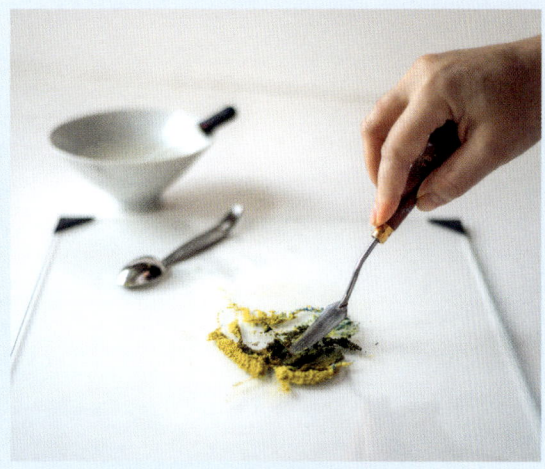

2. On a nonporous surface or slab, measure out roughly 2 tsp pigment. You may choose to mix colors together, or incorporate chalk. Create a divot in the center of your mound of pigment. Add 3.5ml binder.

3. Work the binder into the pigment with a palette knife. Incrementally add more binder by drops as needed, until the consistency is malleable and claylike.

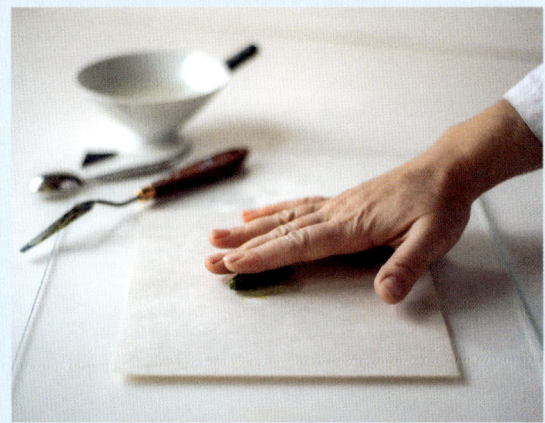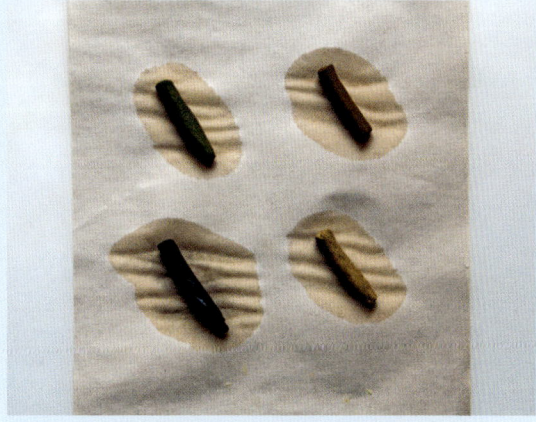

4. Transfer the paste to a piece of newsprint and roll it into a log, holding your fingers parallel to its length. If the pastel cracks apart, return it to the slab and work in a few more drops of binder.

5. Move the finished pastels to a fresh sheet of newsprint to dry for one to two days, weather dependent. When dry, sample your pastels on toothy paper to compare their texture, color, and granularity.

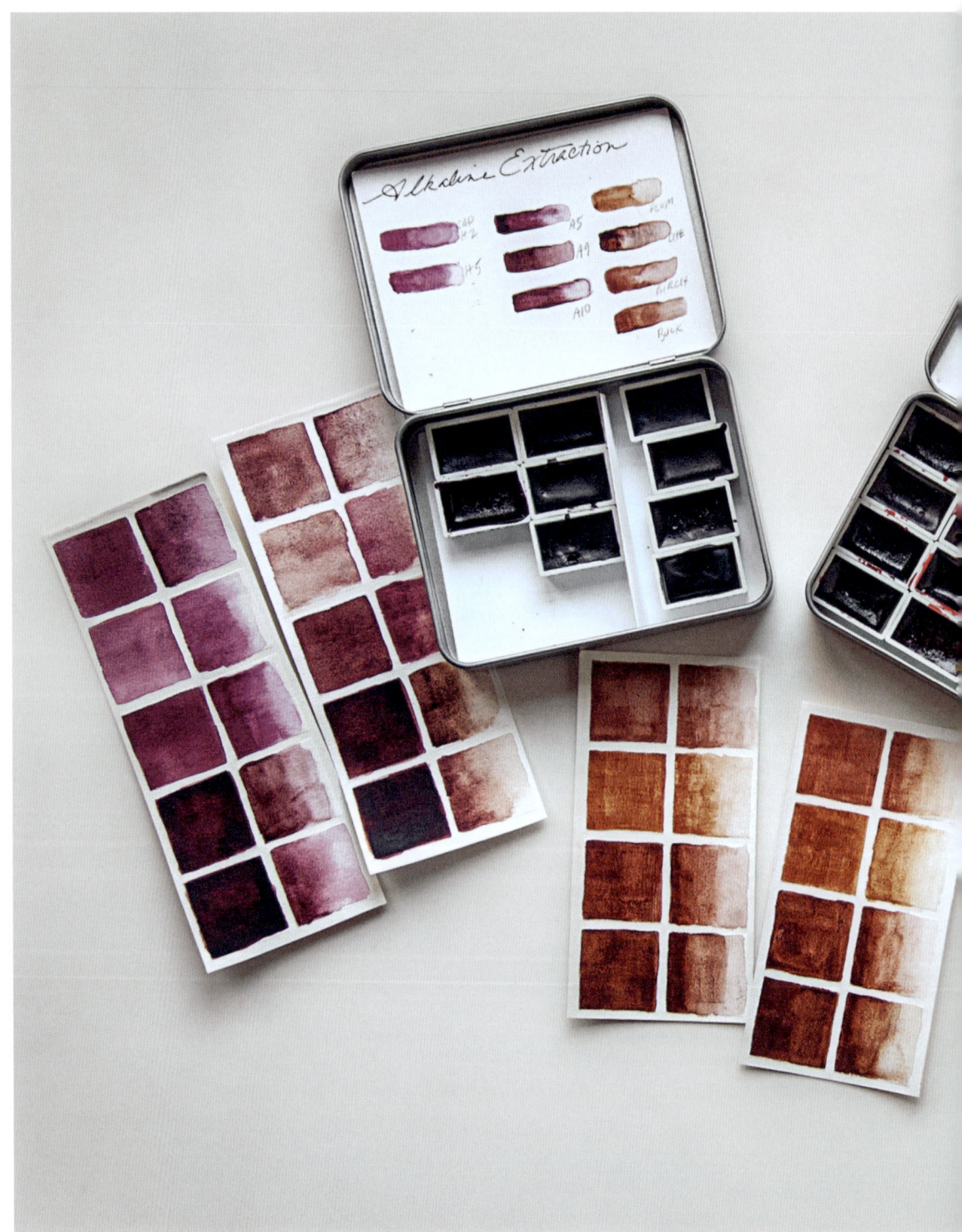

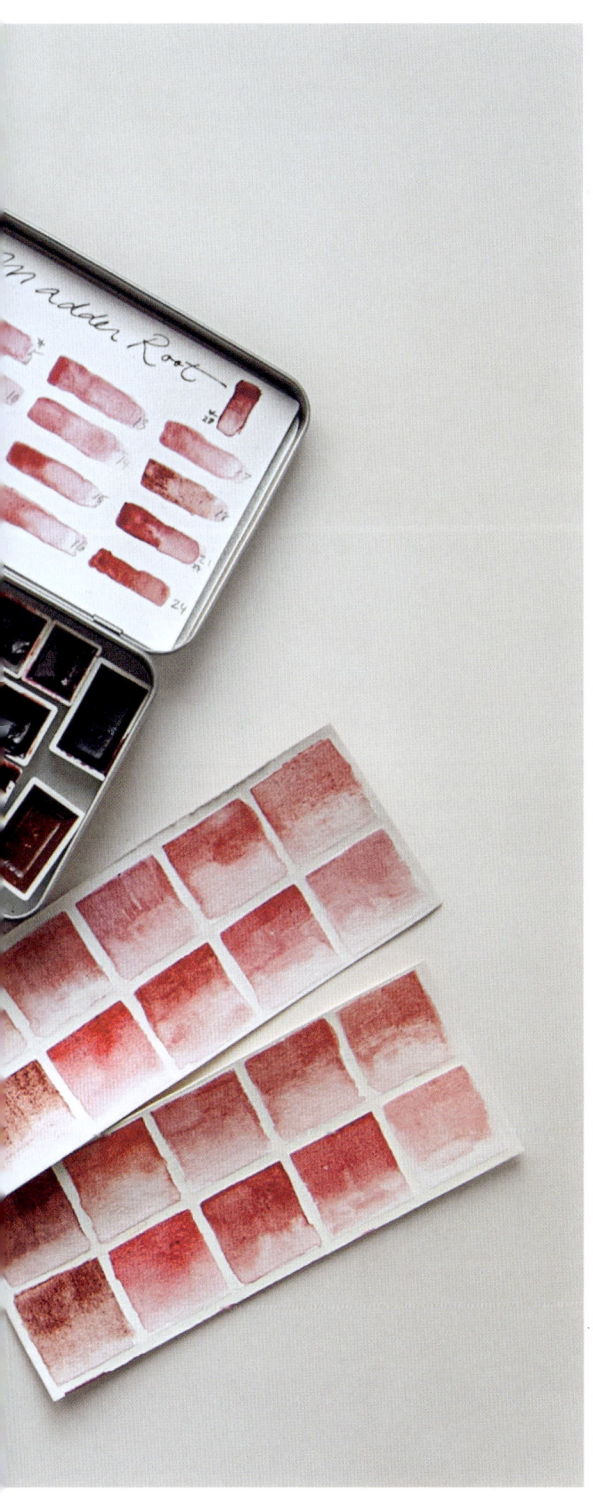

Watercolor

Portable and shelf-stable pans of watercolor are a great way to integrate botanical colors into studio practice and offer a starting point to explore their characteristics. To make this buttery-textured paint, you'll need a muller and a grinding slab. The slab is a hard, nonporous surface, such as marble, porphyry, or glass. Against this slab the flat, broad face of the muller mixes pigment into binder, shearing the agglomerates into minute specks and dispersing them evenly in the binding medium.

Lake pigments made with soda ash are translucent in watercolor medium, while chalk-based lakes dry opaque and may be better termed gouache. When compounding your homemade paint, remember that botanical pigments are compatible with one another, as well as with any other natural or synthetic pigments. They can be mixed on the palette or combined during mulling to create custom paint shades. The process of hand mulling paint is a labor, but it is also an opportunity to observe and familiarize yourself with the characteristics and propensities of your pigments.

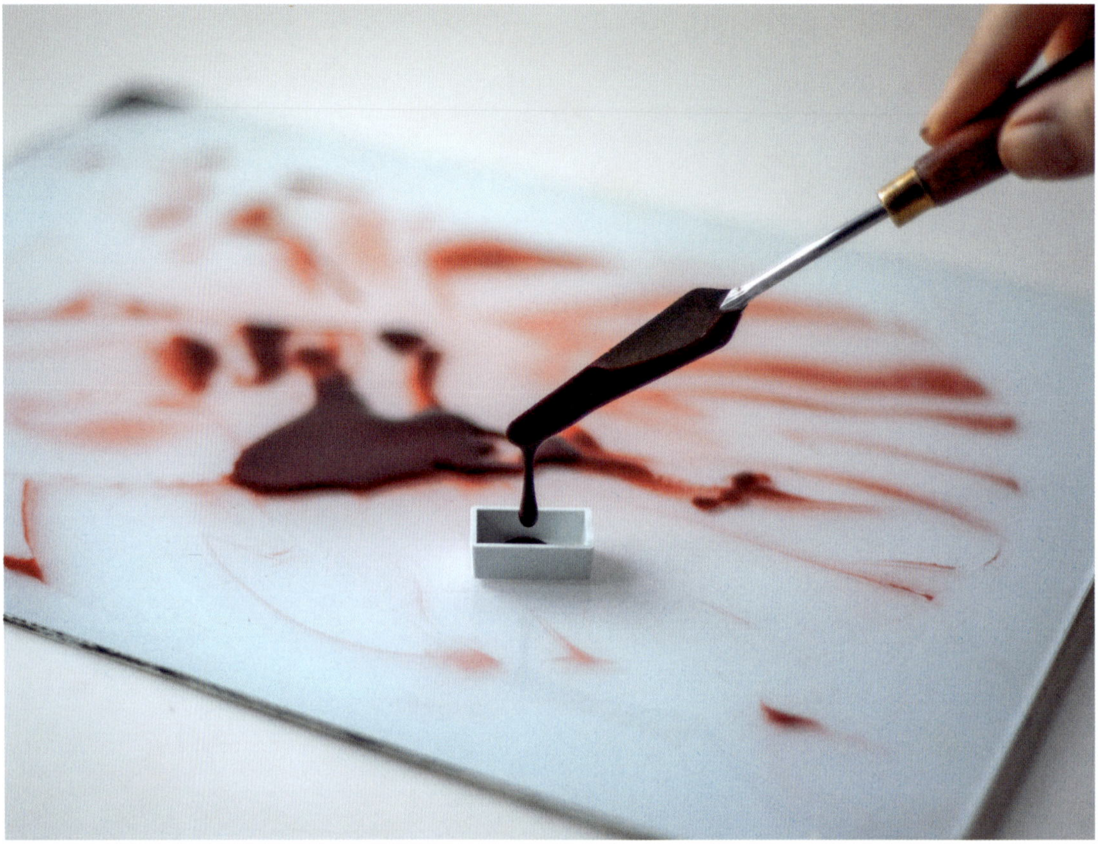

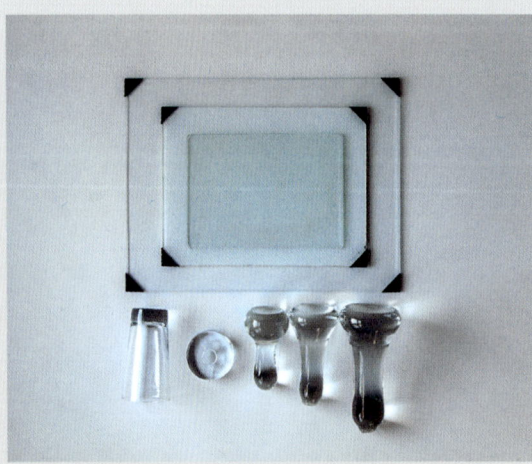

A muller and slab can be sourced from an artists' supplier or scrounged more frugally. Consider a countertop offcut or marble serving platter for a slab; I use a tempered glass paint palette. If your slab is too lightweight to hold its footing, place it on a nonslip mat or damp towel. I have not needed to prepare the slab by roughing up the surface; the pigments themselves perform this service over time. A muller needs hefty weight, a perfectly flat face at least 4cm (1½in) across, and an ergonomic grip. Besides the professional variety available, you might use a glass fermentation weight or thick-bottomed drinking glass.

WATERCOLOR MEDIUM

This watercolor medium is intended for dry pans of paint, whether they're for a travel kit or a gathering of seashells. Gum arabic is the most common binder used in watercolors, along with a few other ingredients that improve their handling qualities and storage. Honey serves as a humectant, attracting moisture so that the paint rewets easily, and clove oil serves as a preservative. Ox gall is a surfactant that allows the paint to flow and mingle on the page (ox gall is an animal product so using it is optional). Feel free to personalize this recipe by varying the ratio of ingredients to suit your preferences. This one makes enough for roughly 20 pans of watercolor.

MATERIALS

· 60g gum arabic, powdered or grains

· 140ml distilled water

· 25ml honey (2 scant tbsp)

· 2 drops clove oil

· 4 drops ox gall, optional

METHOD

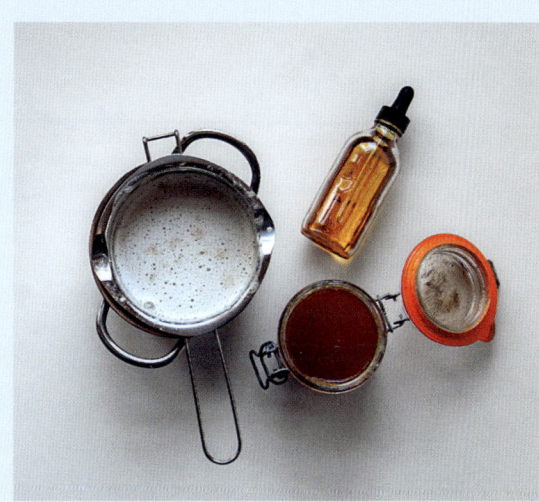

1. Combine gum arabic powder or sap granules with distilled water in the top of a double boiler. Add water to the lower chamber, and simmer for 10 to 20 minutes. Stir the gum arabic mixture in the upper pot occasionally, until dissolved. Remove from heat. If needed, sieve gum arabic to remove grit and twigs. Stir in honey and clove oil. Add ox gall to increase flow (optional). Store watercolor medium in an airtight container in the refrigerator indefinitely.

WATERCOLOR PAINT

This watercolor paint recipe makes one to two full pans.

MATERIALS

· 1 tsp dry pigment, ground through step 2 (page 155)

· Approximately 8 to 12 ml Watercolor Medium (recipe on page 165)

· Distilled water

· Muller and slab

· Palette knife

· Eyedropper

· 2 watercolor pans, bottle caps, seashells, or similar paint pans

METHOD

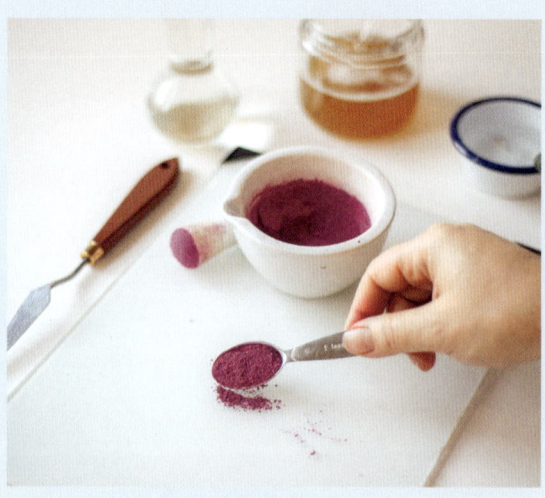

1. Mound 1 tsp ground pigment on your mulling slab.

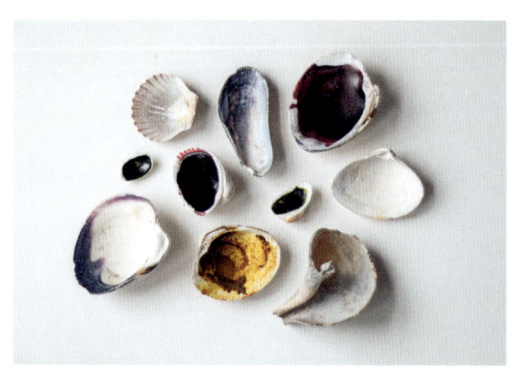

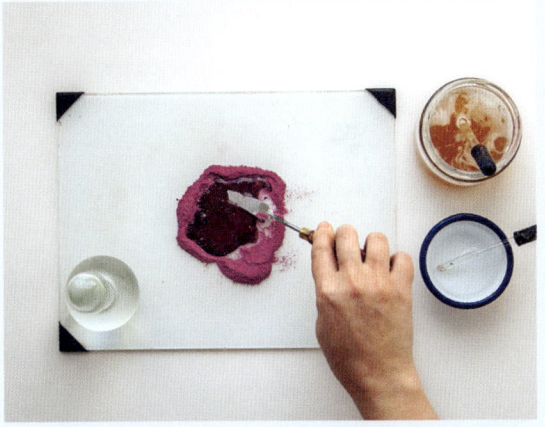

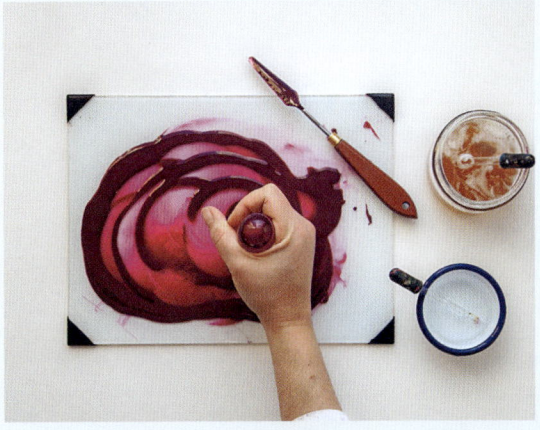

2. Make a well in the pigment and incrementally add watercolor medium, working it into the pigment with a palette knife until the pigment is fully moistened. The amount of medium varies with the specific pigment type; continue adding watercolor medium by drops until the texture is similar to molten chocolate.

3. Use a paint muller to swish the mixture against the slab with a circular or figure-eight motion. There is no need to push down on the muller; it will suction to the slab with the binder, shearing the pigment chunks apart. Add drops of distilled water to keep the paint fluid as the medium begins to dry. Periodically use the palette knife to corral the paint from the edges of both the muller and slab.

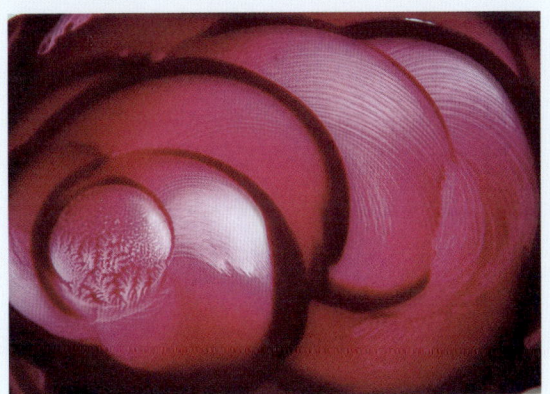

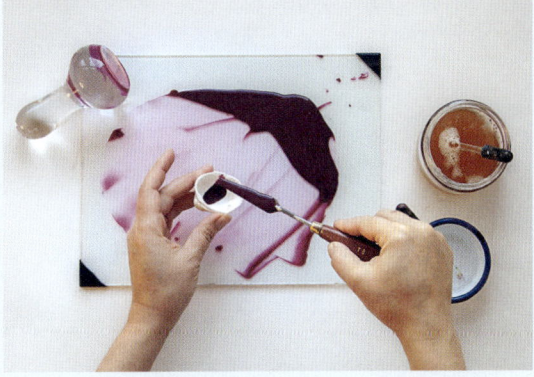

4. Mull until you hear, see, and feel no more graininess, and the paint has a nice body and sheen. To test, paint out a swatch and observe the texture, tackiness, and granularity. If the pigment lifts when dry, it needs more medium. If it fails to dry, it has too much medium and needs more pigment. If the granularity is more than you like, continue mulling to reduce the particle size.

5. Use a palette knife to transfer the paint into pans. Allow to dry.

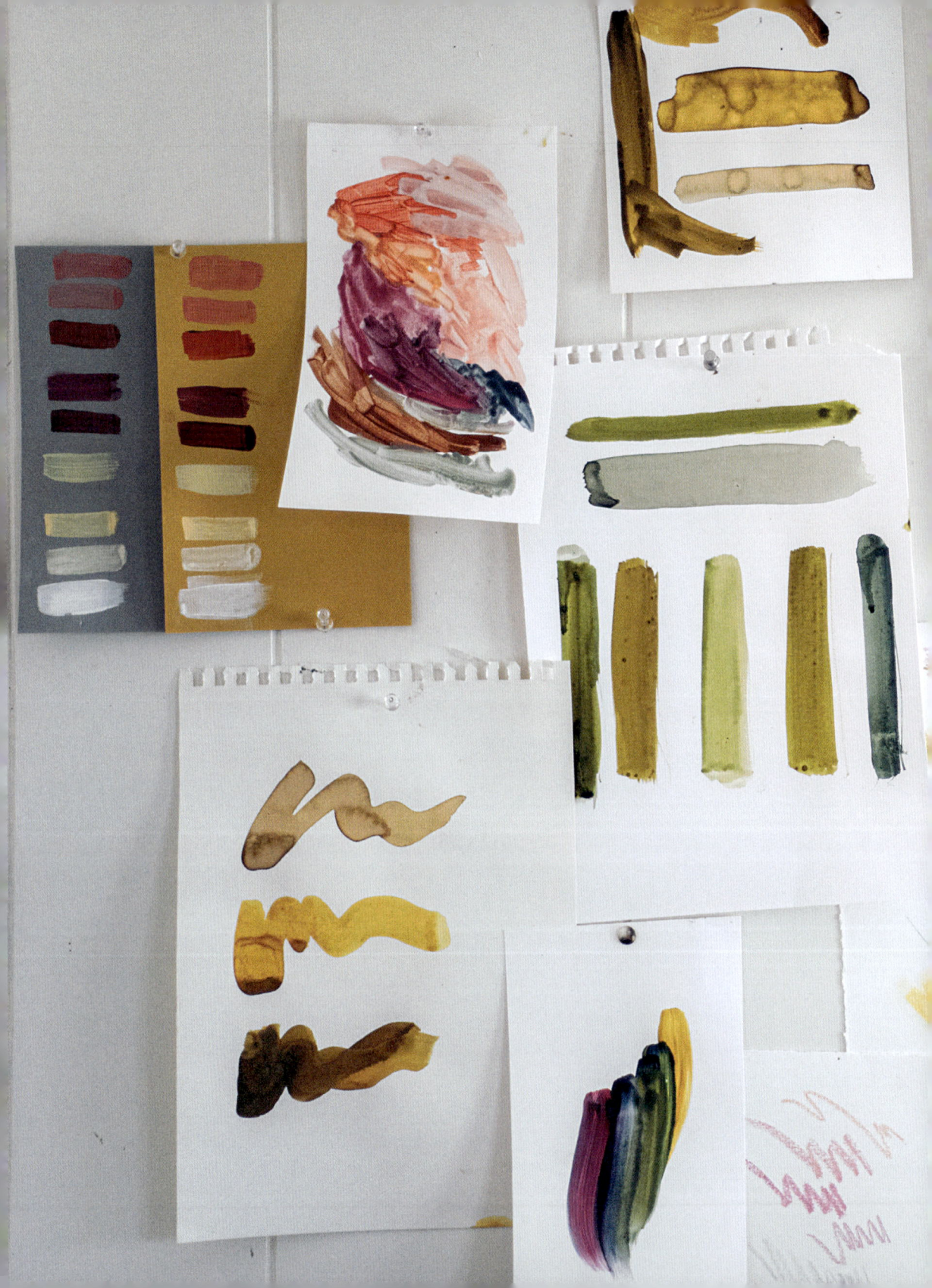